S0-CLG-642

Series on

PROSPERING IN A GLOBAL ECONOMY

National Interests in an Age of Global Technology

Thomas H. Lee and Proctor P. Reid, Editors

Committee on
Engineering as an International Enterprise

NATIONAL ACADEMY OF ENGINEERING

NATIONAL ACADEMY PRESS
WASHINGTON, D.C. 1991

NATIONAL INTERESTS IN AN AGE OF GLOBAL TECHNOLOGY

NATIONAL ACADEMY PRESS • 2101 Constitution Avenue, NW • Washington, DC 20418

NOTICE: The National Academy of Engineering was established in 1964, under the charter of the National Academy of Sciences, as a parallel organization of outstanding engineers. It is autonomous in its administration and in the selection of its members, sharing with the National Academy of Sciences the responsibility for advising the federal government. The National Academy of Engineering also sponsors engineering programs aimed at meeting national needs, encourages education and research, and recognizes the superior achievement of engineers. Dr. Robert M. White is president of the National Academy of Engineering.

This publication has been reviewed by a group other than the authors according to procedures approved by a National Academy of Engineering report review process.

Partial funding for this effort was provided by the Alfred P. Sloan Foundation and the National Academy of Engineering Technology Agenda Program.

Library of Congress Cataloging-in-Publication Data

National interests in an age of global technology / Thomas H. Lee and
Proctor P. Reid, editors.
p. cm.—(Prospering in a global economy)
Includes bibliographical references and index.
ISBN 0-309-04329-8
1. Technology and state—United States. 2. International business
enterprises—United States. I. Lee, Thomas H., 1923- II. Reid, Proctor P. III. Series.
T21.n33 1991
338.97306—dc20 90-20070
CIP

Printed in the United States of America

Committee on Engineering as an International Enterprise

THOMAS H. LEE, *Chairman*, Professor of Electrical Engineering, Massachusetts Institute of Technology
THOMAS D. BARROW, Retired Vice Chairman, Standard Oil Company of Ohio
W. DALE COMPTON, Lillian M. Gilbreth Distinguished Professor of Industrial Engineering, Purdue University
ELMER L. GADEN, JR., Wills Johnson Professor of Chemical Engineering, University of Virginia
DONALD L. HAMMOND, Retired Director, Hewlett-Parkard Laboratories, Hewlett-Packard Company
WILLIAM G. HOWARD, Jr., Senior Fellow, National Academy of Engineering
TREVOR O. JONES, Chairman of the Board, Libby-Owens-Ford Company
MILTON LEVENSON, Executive Engineer, Bechtel Power Corporation
PETER W. LIKINS, President, Lehigh University
EDWARD A. MASON, Retired, Vice President Research, Amoco Corporation
BRIAN H. ROWE, Senior Vice President, GE Aircraft Engines, General Electric Company
WILLIAM J. SPENCER, President and Chief Executive Officer, Sematech
WILLIS S. WHITE, JR., Chairman and Chief Executive Officer, American Electric Power Company

NAE STAFF

PROCTOR P. REID, Study Director, Senior Program Officer
BARBARA L. BECKER, Administrative Assistant
BRUCE R. GUILE, Director, Program Office
H. DALE LANGFORD, Editor
JAMES R. PORTER, NAE Intern
ANNMARIE M. TERRACIANO, Program Assistant

Preface

Since World War II, major transformations of the world's economic, social, and political structures have been taking place on a scale and at a pace unparalleled in history, and this pace has been quickening over the past two decades. The major driving force powering these transformations is technological progress. The unprecedented advances in our understanding of nature are being rapidly and broadly applied and enhanced through technology in industry, agriculture, medicine, and services to meet human needs, wants, and preferences around the world.

The most striking new aspect of these transformations, as compared with past experience, is the speed with which they propagate across national boundaries to reach global dimensions. Scientific, technological, and managerial knowledge diffuse rapidly across these boundaries, enlarging the numbers of nations in which technical competence for engineering and production of a wide range of products may be found. At the same time, the speed and capacity of air transportation bring people, materials, work in progress, and finished goods anywhere in the world in hours. The speed and capacity of satellite and fiber-optic communication and computer networks make possible the closely integrated management of far-flung industrial, financial, and other enterprises and also contribute to tightly linking financial, commodity, and equity markets worldwide.

As a result, the full range of productive activities including research, engineering, production, and marketing in many industrial sectors have increasingly become global in scope, implemented through multinational corporations, foreign direct investments, and international joint ventures. The global span of technology and the global economic activities that result raise new questions about how we think of national interests and national government roles in overseeing and supporting international industrial activity and trade whose domains increasingly overlap with domestic industry and trade. Correspondingly, definitions of domestic and foreign corporations and their relationships to home and host governments in geographic, economic, and political terms have become complex and often difficult to deal with in existing public policy frameworks.

To examine the implications of the rapidly expanding global economy for the engineering enterprise worldwide and especially in the United States, the National Academy of Engineering convened a Committee on Engineering as an International Enterprise. The committee examined in some detail the international aspects of eight specific industrial sectors (included in the appendixes to this report) in addition to reviewing more generally the international factors affecting a wide range of industries. A symposium entitled "National Interests in an Age of Global Technology" held on 4–5 December 1989 at the Arnold and Mabel Beckman Center of the National Academies of Sciences and Engineering provided additional viewpoints and discussions on these subjects. There was also an exchange of information with a contemporaneous study, *The Internationalization of U.S. Manufacturing: Causes and Consequences* (National Academy Press, 1990), conducted under the auspices of the Manufacturing Studies Board of the National Research Council.

Harvey Brooks, Gerald Dinneen , and Alexander Flax provided valuable insights, guidance, and assistance to the committee over the course of the study's development. Bruce Guile, director of the NAE Program Office, contributed valued intellectual stimulus and overall continuity and management support for the project. I wish to thank the study director, Proctor Reid, and the members of the committee for their persistence and hard work in bringing this project to completion, and members of the NAE staff, including Barbara Becker, Dale Langford, James Porter, and Annmarie Terraciano, for their able support.

This report presents some of the more significant information considered by the committee and summarizes the assessments and judgments arrived at in the committee deliberations. The committee has considered the trends and issues that were perceived from the standpoint of engineering and technology in the broad context of public policy—domestic and foreign—and has indicated some ways to help ensure a continuing major role for the United States in a growing and prospering technology-driven world economy.

ROBERT M. WHITE
President
National Academy of Engineering

Contents

Appendixes

Figures and Tables

Figures

Tables

National Interests in an Age of Global Technology

Summary and Recommendations

The rapid globalization of technology during the past two decades has given new meaning to the concept of interdependence for the United States. To compete effectively at home or abroad, many U.S. companies and universities and the nation's technical work force as a whole are becoming increasingly integrated into global networks of research, development, production, and marketing through the expansion of international trade, foreign direct investment, and corporate alliances. These developments have challenged long-standing assumptions regarding the autonomy and supremacy of the U.S. technical enterprise and, in so doing, have fundamentally altered the terms of the traditional competitiveness debate.

Since the mid-1970s, there has been an acceleration of two mutually reinforcing trends—the convergence in technical capabilities of industrialized nations and the global integration of formerly discrete national technical enterprises. The technologically unipolar world of the 1950s and 1960s, dominated by the United States, has given way in the past decade and a half to a world in which technical competence and resources are much more dispersed among a number of industrialized and industrializing countries. International comparisons of patenting, R&D spending and personnel, high-tech trade and production, and foreign direct investment since the mid-1970s all evidence this trend (see Chapter 1, pp. 14–23).

In concert with this profound change in the global distribution of technical capabilities, the organization of the advanced technical activities of corporations has become increasingly transnational. From the end of World War II to the early 1970s, the internationalization of production was driven primarily by U.S. foreign direct investment. During this period, production in many industries became increasingly multinational or global, but advanced technical activities such as research and development remained predominantly "national," that is, concentrated in the major corporations' home countries. **During the last decade and a half, however, a new model of internationalization has emerged, characterized by the rapid growth of non-U.S. foreign direct investment and a proliferation of transnational corporate alliances. The globalization of production in the 1980s and beyond encompasses the full spectrum of corporate technical activities** (see Chapter 1, pp. 23–25).

Responding to the challenges and opportunities of increased global competition, shorter product cycles, national "managed trade" policies of varying scope, wider markets, and a growing number of globally dispersed sources of new technology and technical competence, transnational companies in many industries have reorganized their technical activities on a global basis. U.S.-based corporations have taken the lead in decentralizing and dispersing their own advanced technical activities internationally, developing and acquiring more of their technology abroad. During the 1980s, transnational corporate alliances, a majority of them involving U.S. corporations, emerged as a major vehicle for gaining access to foreign markets and technology. Although U.S.-based multinationals have been forerunners of a trend, they are not alone. As their technical prowess and foreign direct investments have expanded, a growing number of foreign corporations have also begun to reorganize their advanced technical activities more internationally and to assume a more active role in the creation of transnational technical alliances (see Chapter 1, pp. 25–35).

The convergence of national technical capabilities and the globalization of advanced technical activities at the hands of multinational corporations underline the growing economic and technical interdependence of nations. **The committee is convinced that the globalization of R&D, production, investment, markets, and technology is a positive trend for both the United States and the rest of the world, although it is not without its problems**. To be sure, the economic, technical, and political imperatives of globalization have created an international environment in which technical capabilities that many deem essential to a nation's continued prosperity and security can be eroded swiftly by intense competition from abroad. Nevertheless, the committee agrees that the benefits and opportunities provided by the globalization trend outweigh any adjustment costs that follow in its wake. Not only does the globalization process accelerate transnational integration and cross-fertilization in engineering, technology, and management, it also promises to enhance the diversity and depth of world engineering and scientific resources and thereby stimulate economic growth and technology development. Most important, the globalization of technical activities cannot be reversed or significantly impeded by national governments without inflicting high costs on their citizens (see Chapter 2, pp. 45–47, 52).

As the past decade has made clear, however, increased international interdependence has not diminished the competitive pursuit of economic and technical advantage by nations. Nor have the benefits (real and potential) of globalization dissuaded governments from pursuing policies that run counter to the larger trends. Governments worldwide have long intervened in their domestic economies to increase the productivity and international competitiveness of firms operating, if not originating, within their borders. However, as more countries have recognized the importance of technical advance for economic growth and competitiveness, governments have focused more on creating a domestic environment conducive to developing, applying, and diffusing advanced technology for commercial advantage. In this quest for economic advantage, nations rely on a range of policy instruments. Some of these are more interventionist, such as "managed trade," domestic content legislation, or "closed" national technology development initiatives; others are more market-oriented, such as deregulation or investments in education and economic infrastructure (see Chapter 2, pp. 47–52; Chapter 4, pp. 72–73).

This new technology-oriented competition among nations is greatly complicated by the blurring of corporate nationalities and the lack of internationally accepted rules of behavior for companies and their home and host governments. As private corporations, which have long been viewed as the mainstays of a nation's commercial technical enterprise, have become more cosmopolitan in outlook and conduct, the relationship between corporate interests and national interests has grown increasingly complex. It is a relationship that requires more deliberate and careful examination. Indeed, the definition of what constitutes a "domestic" or a "foreign" corporation and the nature of "corporate citizenship" more generally have become more and more vexing issues for public policymakers as the technical activities and resource base of a growing number of corporations become increasingly distributed internationally.

Similarly, the emerging global economic and technical enterprise challenges long-standing assumptions regarding the relatively neat dichotomy of domestic and international policy areas related to national competitiveness. To deal effectively with the domestic and international political friction that accompanies the globalization trend, national governments are being called upon to negotiate internationally areas of public policy traditionally viewed as exclusively matters of domestic concern (see Chapter 4, pp. 73–74).

The changing character of competition among corporations and the competitive pursuit of economic advantage among nations in an age of increasing international technical interdependence pose several major challenges for the United States. More than any other advanced industrialized country, the United States has long considered itself technologically self-sufficient and has relied heavily on the technical superiority of its indigenous companies to sustain an advantageous position in the world economy. Although the United States remains the world's most technologically self-sufficient country, its economic prosperity and technical dynamism have already become highly dependent on foreign technology, capital, and markets and are likely to become more so in the coming decades. **Indeed, the technical and economic vitality of the United States depends increasingly on the ability of companies operating within its borders to harness and exploit globally dispersed resources and technical capabilities rapidly and effectively** (see Chapter 1, pp. 25–26).

In addition, **the rapid growth of technical competence beyond U.S. borders has made it increasingly difficult for U.S.-based companies to derive sustained competitive advantages from superior research capabilities alone.** As foreign nations and companies have acquired greater technical capabilities, new knowledge or basic research increasingly has become a "global public good," impossible to bottle up within any one nation's borders, and easily accessible to any and all takers. To prosper in this environment, it is becoming imperative that U.S.-based corporations compete effectively at every step along the way in the conversion of scientific discoveries into commercial services or products. Although the United States is renowned for the strength and breadth of its research enterprise, a growing number of U.S.-based companies appear to be at a disadvantage in relation to their Japanese and other foreign competitors in the downstream technical activities critical to leveraging technology for commercial advantage—technology development, acquisition, adaptation, and diffusion (see Chapter 1, pp. 29–35; Chapter 3; Appendix A, pp. 89–135).

Drawing on a series of industry case studies, the proceedings of committee meetings and a major symposium,[1] and the views of many knowledgeable representatives from government, industry, and academe in North America, Western Europe, and Asia, this study argues for more explicit recognition of the emerging global technical enterprise and its profound implications for private strategies and public policies. In the judgment of the committee, the national and international policy debate must be recast to square with the realities of global technical convergence and interdependence.

CAPTURING THE BENEFITS OF GLOBAL TECHNICAL ADVANCE

The highest priority for strengthening the technical foundations and thereby the long-term wealth-generating capacity of the U.S. economy must be to make the United States a more attractive and advantageous place for individuals, companies, and other institutional entities,

regardless of national origin, to conduct the full complement of technical activities critical to the nation's long-term prosperity and security. To accomplish this, the United States must develop the necessary human, financial, physical, regulatory, and institutional infrastructures to compare more advantageously with other nations in attracting the technical, managerial, and financial resources of globally active private corporations or individuals. This is the single most important conclusion of the study.

Clearly, all sectors of U.S. society—industry, government, and both basic and higher education—have important roles to play in this effort. The committee has focused primarily on public policy implications, but it does not believe that public policies are the only or even the most important determinants of national or corporate technical strength and competitiveness.[2] Rather, the study's public policy focus has been shaped by the fact that the public sector is groping to formulate and implement a national agenda that can address the imperatives of a highly integrated global economic and technical order.

The government must take action on many fronts to strengthen the foundations of the U.S. technical enterprise—the nation's work force, its social capital (i.e., educational system and public infrastructure), as well as its fiscal and regulatory environment. **Above all, state and federal policymakers must work together with corporate and academic leaders to develop a broad national consensus regarding the need to improve technology development, adoption, adaptation, and diffusion throughout the U.S. industrial economy. This consensus, in concert with other national policies, can provide the necessary impetus, coherence, and operational guidelines for the many diverse private and public policy actions required to meet the challenges of globalization.**

DOMESTIC POLICY DIRECTIONS

Among the greatest comparative strengths of the nation's technical enterprise are its research capabilities, its system of advanced technical education, its large pool of elite technical talent, and its extensive, sophisticated information technology infrastructure. These comparative advantages find expression in continuing U.S. commercial leadership in highly science-intensive industries or industries in the infancy of their technology life cycle. Moreover, the nation's extensive research enterprise provides the human and intellectual resources for much of U.S. high-technology industry, attracts foreign talent and investment to the country, and benefits U.S. citizens in many other ways. In the opinion of the committee, it is imperative that the United States continue to build on these comparative strengths (see Chapter 3, pp. 54–61; Chapter 4, pp. 77–78).

The recent intensity of global competition and the pace of technical advance have underlined the growing importance of synergies between basic research and downstream technical activities such as product and process design, development, and production in many industries. Nevertheless, the past two decades have also demonstrated that as new knowledge flows more freely across national borders, the ability of a nation or a firm to exploit research results for commercial advantage depends increasingly on mastery of those downstream technical activities.

This trend is particularly troublesome for the United States, which continues to harbor the world's most extensive and productive basic research enterprise even as the ability of many U.S.-based industries to adopt and adapt technology for commercial gain appears to have declined relative to other nations. The inability of many U.S.-based industries to derive what many consider a fair share of commercial benefits from an increasingly global technology base underlines the need for U.S. educators, industrialists, and policymakers to direct greater attention and resources toward "relearning" these vital activities—competencies closely associated with the production of goods and services in which the United States excelled from the late 1800s well into the mid-1900s (see Chapter 1, pp. 29–35; Chapter 3, pp. 61–67; Appendix A, pp. 89–135).

The committee views the following domestic policy directions as essential elements of a more comprehensive technology strategy for the United States.

- **Policymakers should expand support for initiatives at the federal, regional, and state levels to enhance the adoption, adaptation, and diffusion of technology and related know-how.** Current federal science and technology policies are targeted primarily on basic research and "mission-oriented" technology development related to national defense, public health, and space exploration. While reinforcing the current U.S. comparative advantage in certain highly science-intensive or "emerging technology" industries, this policy orientation essentially neglects national vulnerabilities in technology adoption, adaptation, and diffusion, which are equally critical to national economic growth and industrial competitiveness.

Recent U.S. experience has demonstrated that low-cost, pragmatic initiatives at the state, regional, or federal level can effectively support private-sector progress in these areas. The National Science Foundation's Engineering Research Centers, the National Institute of Standards and Technology's Centers for Manufacturing Technology, Ohio's Thomas Edison Program, Pennsylvania's Ben Franklin Partnership Program, the Southern Technology Council, and the Industrial Technology Institute are promising means for providing public support for a diverse set of initiatives and selectively broadening the application of those that prove most successful (see National Academy of Engineering, 1990; National Governors' Association, 1988; National Research Council, 1990b; Pennsylvania Department of Commerce, 1988)[3] (Chapter 4, pp. 78–79).

- **U.S. public policy should acknowledge the need for a stronger public role in support of generic technologies and establish credible mechanisms for translating this commitment in principle into specific actions.** There is a need for the United States to develop more focused national or regional infrastructures for supporting the development and diffusion of commercially significant generic technologies. Such technologies involve concepts of design, fabrication, and quality control applicable to a class of products, for which (a) the anticipated returns from development and commercialization cannot justify the expense and risk of investment by single firms or joint ventures; and (b) the returns to the economy and society as a whole warrant investment by the federal government. In addition, there may be areas in which national military strategic considerations make loss of U.S. technology position or of market share unacceptable.

Promotion of commercially significant generic technologies need not require major investments in research and development programs. Indeed, obstacles to the diffusion of such technologies may be more important than any obstacle to their development. To be sure, significant public and private investment may be required in certain cases, as in the development of a new generation of semiconductors, when the cost of technological advance is so high, the time scale of technology development is very long, and the ability of any one firm to benefit from such large investments is so low or unpredictable that no firm is willing to take the risk. For other generic technologies, however, development costs may not be high—or the technology may already be available—yet there may be serious economic, regulatory, or societal obstacles to the adoption, adaptation, and diffusion of the technology either within or across industries. For example, "total quality control" methods, computer-aided design, advanced construction techniques, and just-in-time production systems are all generic technologies that might fall into this category.

There is, at present, considerable debate regarding the proper government role in support of generic technologies. In the opinion of the committee, the primary roles of government should be as convener and catalyst of such activities undertaken in the private sector and may also involve harnessing the technical resources of the nation's federal laboratories more directly in support of high-cost, high-risk, nonappropriable generic technology development. In some cases this may involve federal matching of a significant amount of private funding. However, in most instances the government should be prepared to serve as the "pathfinder," providing more indirect fiscal or regulatory support to private-sector participants.

Ultimately any effort to provide government support for the development and diffusion of generic technology in the United States will depend on the credibility of the public and private institutional mechanisms designated to assess and identify those technologies most in need of attention and to chart an appropriate policy response. The committee notes that there have been several attempts by federal agencies to identify "critical" technologies in recent months, most notably by the departments of Commerce (1990) and Defense (1990). The mixed reception of these efforts in the U.S. policy community, however, underlines the need for institutions that assume this charge to be perceived as technically expert, responsive to the interests of all U.S. citizens—consumers, producers, and suppliers—and predisposed to operate in a manner consistent with emerging global economic and technological realities (see Chapter 4, pp. 79–81).

• **Public policy initiatives to strengthen the national technology and industry base should be guided by the extent to which a corporation genuinely contributes to the national economy. With rare exception, such policies should not discriminate among corporations on the basis of nationality of ownership or incorporation, provided there is sufficient reciprocity in the large.** Public sector assistance to, or collaboration with, private corporations (domestic or foreign) in pursuit of national objectives should be governed by common standards for the corporate role in the U.S. economy. It is entirely appropriate that policymakers charged with advancing the interests of all U.S. citizens should develop criteria consistent with that charge regarding corporate participation in any venture involving public funds or legal exemptions. In a global economy with globally active corporations, however, corporate nationality is a poor measure of a firm's real or potential contribution to U.S. national interests. There may be circumstances in which the U.S. government should discriminate against foreign-owned firms temporarily to achieve reciprocal equitable "national treatment" of U.S. companies doing business overseas or to safeguard national security. However, nondiscrimination with regard to corporate nationality should remain a key principle of U.S. public policy (see Chapter 4, pp. 81–82).

• **State and federal governments should redouble their efforts to modernize and strengthen the nation's work force and public infrastructure and to encourage continuous modernization of plant and equipment in private industry.** The continuing globalization of technology and the resulting intensification of competition among firms and nations impart an increasing sense of urgency to this familiar recommendation (see Council on Competitiveness, 1988; National Academy of Engineering, 1988a, 1988b; President's Commission on Industrial Competitiveness, 1985). New technology by itself will not generate the wealth or productivity increases necessary to increase the standard of living of U.S. citizens and strengthen U.S. national competitiveness. These objectives demand that the United States devote greater attention to the social and human capital that supports the technological capabilities and commercial vitality of corporations based or operating in the United States. Public sector investment in the nation's educational system and physical infrastructure is vital. Government should create a fiscal and regulatory environment that will encourage private industry to invest in plant, equipment, and organizational learning that will enable it to develop, adopt, and adapt technology more effectively for commercial gain (see Chapter 3, pp. 61–66; Chapter 4, pp. 82–83).

• **Government should devote greater attention to the technological dimensions of international trade, investment, competition, and other critical issues not traditionally associated with science and technology concerns. To this end, government should seek to cultivate greater technical expertise in agencies responsible for domestic and international economic policy, and to improve interagency communication and coordination regarding science and technology issues.** The development and commercialization of technology are not a discrete policy issue but an integral part of many broader areas of domestic and foreign policy. Until recently, there has been insufficient appreciation of implications for science and technology policy initiatives across agencies. There has been even less communication and cooperation among those responsible for formulating and implementing domestic and foreign policies that bear on the health of the nation's commercial technology base. This situation argues for expanding recruitment of technically competent personnel by agencies that formulate and implement domestic and international economic policy and also points up the need for greater organizational focus at the national level on the policies affecting commercial development and application of technology.

The committee notes with guarded optimism the positive steps by the current administration to provide more organizational focus through the President's Science and Technology Adviser, recently elevated to the position of Assistant to the President, the President's Council of Advisers on Science and Technology, the Office of Science and Technology Policy, the newly created Office of Technology Policy in the Department of Commerce, and Commerce's National Institute of Standards and Technology. These bodies clearly have the potential for improving intragovernmental communication and coordination across a range of domestic and international policy areas related to technology and economics. Ultimately, it is of secondary importance whether the necessary organizational focus is located in a single independent agency (existing or to be created) or finds expression in more institutionalized interaction among the many agencies and committees that currently influence the nation's technology base. What is critical is that those seeking to develop greater organizational focus acknowledge the growing synergies between what have traditionally been viewed as discrete policy areas (see Chapter 4, pp. 83–84).

INTERNATIONAL POLICY DIRECTIONS

The increasingly global character of corporate technical activities has made it essential that policies aimed at developing and better managing the nation's technical endowments be outward looking—consistent with an international policy framework that fosters and structures technological competition, cooperation, and exchange among nations and firms. Ultimately, the nation's ability to capture a fair share of the benefits of the global technical enterprise will depend primarily on the extent to which private corporations operating within its borders seize the opportunities presented by the emerging global technology base. Their success or failure, however, will be conditioned by the extent to which U.S. policymakers recognize the interdependence of domestic and international policies that influence technology development, diffusion, and commercialization.

In foreign relations, there are a number of things the United States can do to complement domestic efforts, promote more reciprocal technical exchange, and attenuate tendencies toward technology-based protectionism. There is an obvious need for continued efforts to liberalize world trade as well as greater public and private involvement in the international standards-setting process, and in the quest for a more effective international intellectual property rights regime. Yet, these high-profile concerns are distracting policymakers from equally important issues raised by the rapid growth of foreign direct investment and transnational corporate alliances and technical networks over the past decade. From the perspective of the U.S. technical enterprise, the most important challenges to U.S. foreign economic policy relate to national disparities in the treatment of foreign direct investment and competition policy.

- **The United States should seek to forge multilateral consensus regarding the mutual obligations of multinational corporations and their home and host governments.** In an effort to improve the nation's trade balance, and to respond more forcefully to a lack of reciprocity overseas, some recent U.S. legislation raises issues related to the free flow of foreign direct investment and the treatment of subsidiaries of foreign-owned corporations.[4] The rapidly increasing foreign penetration of the U.S. economy in the past two decades has generated a great deal of concern among many segments of the American electorate. Furthermore, the discriminatory treatment of U.S.-owned corporations appears to be a fact of life in Japan and to be increasing in Western Europe as the countries of the European Community search for ways to come to terms with intensifying global competition and the consequences of a new round of economic integration within the European Community (EC 1992). Nevertheless, discriminatory policies are not consistent with global economic and technological realities and may be counterproductive in the long run. In the committee's judgment, such policies would be detrimental to U.S. national interests. Given the extent of U.S. global technological interdependence, and the many contributions of the U.S. subsidiaries of foreign firms to the U.S. economy and technical enterprise, it is particularly important that the U.S. market remain open to foreign direct investment and that, as far as possible, such open-market policies be reciprocal.

The committee recognizes that there are many troubling issues raised by the recent growth in foreign control over U.S. industrial assets and the extent to which foreign multinationals draw upon the U.S. research enterprise. It does suggest, however, that it is time for a more multilateral approach to foreign direct investment—an approach that acknowledges the pervasive character and positive contributions of foreign direct investment in an effort to arrive at mutually beneficial "rules of the game" for both transnational corporations and their home and host countries. Good corporate citizenship is becoming ever harder to define as the operations of U.S. and foreign-owned firms become increasingly transnational. An aggressive U.S. effort to forge multilateral consensus regarding the mutual obligations of multinational corporations and their host governments would do much to reduce tendencies toward technology-oriented protectionism worldwide as well as expand international technology exchange (see Chapter 1, pp. 25–26; Chapter 2, pp. 48–49; Chapter 4, pp. 84–85).

- **U.S. policymakers should strive for greater uniformity in antitrust policy at the international level.** There is mounting pressure on policymakers throughout the industrialized world to reinterpret national antitrust law or competition policy to fit the realities of global competition and avoid disadvantaging their indigenous firms in the global marketplace. Nevertheless, in the context of the current surge of foreign direct investment and the proliferation of transnational corporate alliances and mergers, often in already highly concentrated industries, unilateral approaches to antitrust regulation pose two major hazards.

On the one hand, relaxation of antitrust requirements by the world's leading economies may increase opportunities for monopoly abuse in certain industries and actually impede technological advance. Although there is little evidence of anticompetitive behavior in manufacturing and service industries at the international level, alliances among former competitors in certain industries and the rising barriers to market entry as a result of the spiraling cost of technical advance create an environment in which anticompetitive behavior is increasingly credible. Despite the possible benefits of interfirm collaboration, it is essential to uphold competition as a major driver for technological advance and structural adjustment.

On the other hand, there is some evidence that national competition or antitrust laws may impede cross-border mergers and acquisitions that do *not* undermine competition. Such policy-induced obstacles to international competition may also impede technological advance and economic growth.

Both the danger of anticompetitive abuse by global companies and the costs of "protectionist" antitrust regulation emphasize a growing need for greater international cooperation in antitrust policy. Multilateral discussion of this issue within the General Agreement on Tariffs and Trade and the Organization for Economic Cooperation and Development warrants greater attention and resolve from all industrialized nations, including the United States (see Chapter 2, pp. 51–52; Chapter 4, pp. 85–86).

NOTES

1. "National Interests in an Age of Global Technology," sponsored by the National Academy of Engineering, 4–5 December 1989, in Irvine, California.
2. For more extensive discussion of the implications of globalization for corporate strategy, see the recent report on the internationalization of U.S. manufacturing issued by the National Research Council (1990a).
3. The Southern Technology Council is based in Research Triangle Park, North Carolina; the Industrial Technology Institute is based in Ann Arbor, Michigan.
4. Consider, for example, the Exon-Florio amendment to the Omnibus Trade and Competitiveness Act of 1988, or the spate of bills currently pending in Congress, including the American Technology Preeminence Act (H.R. 4329), Technology Corporation Act of 1990, and others that seek to spell out in legislation specific "special" requirements for foreign-owned or foreign-controlled firms' participation in publicly funded research and development initiatives.

REFERENCES

Council on Competitiveness. 1988. Picking Up the Pace: The Commercial Challenge to American Innovation. Washington, D.C.: Council on Competitiveness.

National Academy of Engineering. 1988a. Focus on the Future: A National Action Plan for Career-Long Education for Engineers. Washington, D.C.: National Academy Press.

National Academy of Engineering. 1988b. The Technological Dimensions of International Competitiveness. Committee on Technology Issues That Impact International Competitiveness. Washington, D.C.

National Academy of Engineering. 1990. Assessment of the National Science Foundation's Engineering Research Centers Program. Washington, D.C.: National Academy Press.

National Governors' Association. 1988. State-Supported SBIR [Small-Business-Innovation-Research] Programs and Related State Technology Programs. Marianne K. Clarke, Center for Policy Research and Analysis. Washington, D.C.: National Governors' Association.

National Research Council. 1990a. The Internationalization of U.S. Manufacturing: Causes and Consequences. Manufacturing Studies Board. Commission on Engineering and Technical Systems. Washington, D.C.: National Academy Press.

National Research Council. 1990b. Ohio's Thomas Edison Centers: A 1990 Review. Commission on Engineering and Technical Systems. Washington, D.C.: National Academy Press.

Pennsylvania Department of Commerce. 1988. Ben Franklin Partnership: Challenge Grant Program for Technological Innovation—Five Year Report. Board of the Ben Franklin Partnership Fund. Harrisburg, Pa.: Pennsylvania Department of Commerce.

President's Commission on Industrial Competitiveness. 1985. Global Competition: The New Reality. Washington, D.C.: U.S. Government Printing Office.

U.S. Department of Commerce. 1990. Emerging Technologies: A Survey of Technical and Economic Opportunities. Office of Technology Administration.

U.S. Department of Defense. 1990. Critical Technologies Plan. Prepared for the Committees on Armed Services, United States Congress. March 15.

1

The Emerging Global Technical Enterprise

The last two decades represent a watershed in the global distribution and organization of technological activities. Since the mid-1970s, there has been an acceleration of two long-standing, mutually reinforcing trends—the convergence in technical capabilities of industrialized nations and the global integration of national technology markets. The virtual elimination of the twentieth century "technology gap" between the United States and its major trading partners in Western Europe and Japan and the rapid growth in technical competence of an expanding group of newly industrialized nations have greatly intensified international technical and commercial competition. Global competition and the advance of technical convergence, in turn, have been accompanied by a surge in international foreign direct investment and a proliferation of transnational corporate networks and technical alliances that have accelerated the integration of formerly relatively discrete national technology markets and industrial activities.

CONVERGENCE IN TECHNICAL CAPABILITIES OF INDUSTRIALIZED NATIONS

Since the 1950s, most of the industrialized and industrializing nations of Europe and Asia have made steady progress toward closing the huge technology and productivity gaps that opened between them and the United States during the first half of the twentieth century.[1] By the late-1980s, America's major industrialized competitors, led by Japan, had greatly expanded their respective national technical capabilities, all but eliminated the U.S. margin in manufacturing productivity, and achieved rough technical and commercial parity with the United States across a range of industries and technologies.

It is easy to challenge the validity or accuracy of any one indicator of change in the relative technological capabilities of nations. Indeed, there is little consensus regarding the significance of comparative patent data as there is concerning the accuracy and meaning of international comparisons of R&D spending or scientific and engineering personnel. Yet by drawing on a range of indicators that include measures of a nation's technical inputs (R&D spending, technical work force) and outputs (patents, high-tech trade and production), as well as measures of the relative efficiency with which these technical resources are employed (productivity), it is possible to provide a multidimensional overview of recent trends in the global balance of commercial technical power.

From the perspective of inputs, the United States continues to boast the world's largest R&D budget as well as the largest national contingent of engineers and scientists. Yet America's competitors have made significant strides during the last two decades, greatly narrowing the differential in human and capital resources.[2] A comparison of recent changes in the ratio of R&D personnel per 10,000 employees for The Group of Five (G-5) economies—Federal Republic of Germany, France, Japan, United Kingdom, and the United States—illustrates this point quite elegantly (Figure 1.1).

Since the late 1960s, the Western Europeans and the Japanese have come

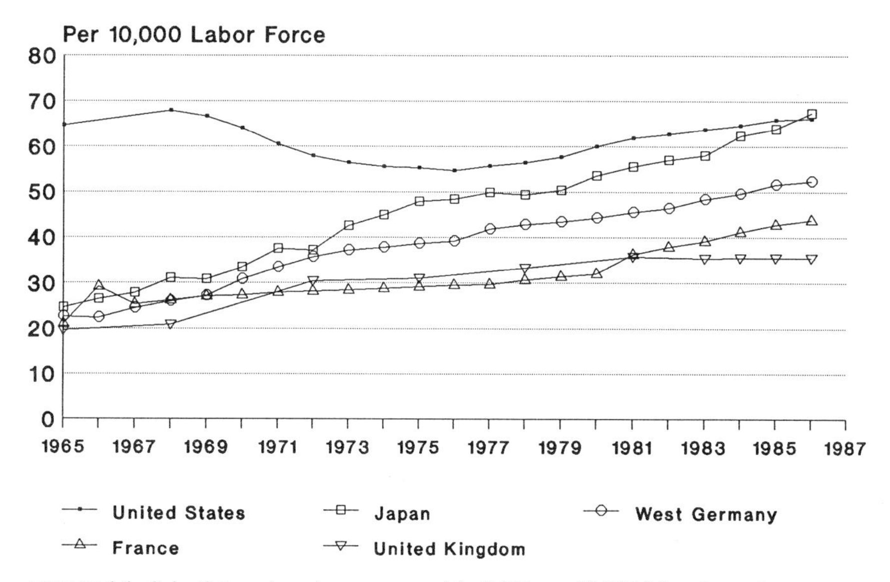

FIGURE 1.1 Scientists and engineers engaged in R&D per 10,000 labor force, by country: 1965–1986. SOURCE: National Science Foundation (1988, p. 38).

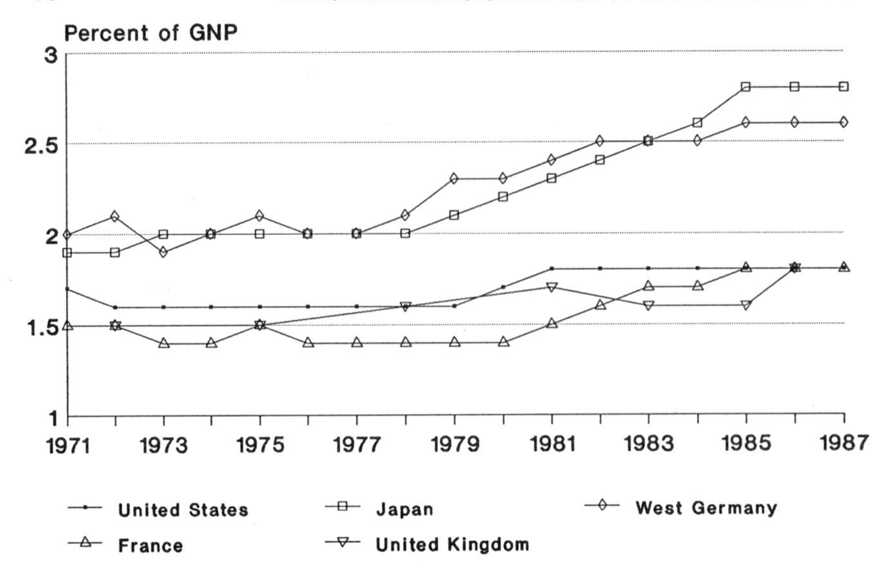

FIGURE 1.2 Estimated nondefense R&D expenditures as a percent of GNP, by country: 1971–1987. French data are based on GDP; consequently, percentages may be slightly overstated to GNP. Foreign currency conversions to U.S. dollars are calculated based on OECD purchasing power parity exchange rates. Constant 1982 dollars are based on U.S. Department of Commerce GNP implicit price deflators. SOURCE: National Science Foundation (1988, p. 8).

a long way toward closing the gap with the United States. Although most of the convergence occurred during the 1970s, Japan continued to increase its ratio during the 1980s, surpassing the U.S. ratio in 1986. Moreover, given the fact that nearly a fifth of total U.S. R&D personnel are engaged in defense-related work, that is, work of limited commercial relevance, the importance of the U.S. absolute margin in R&D personnel is clearly diminished.[3]

A similar picture emerges from a comparison of nondefense R&D spending as a percentage of GNP for these five countries (Figure 1.2). The United States has historically channeled a significantly larger share of its total R&D funds to defense purposes than its trading partners, anywhere from a quarter to a third of U.S. R&D expenditures in recent decades.[4] However, from the mid-1970s to the late-1980s, a period of greatly intensified global industrial competition during which the relevance of defense R&D to commercial applications has declined markedly, growth in the ratio of nondefense R&D to GNP for the United States remained relatively flat while that for Japan, the Federal Republic of Germany, and, to a lesser extent, France, experienced significant growth.

America's major competitors have also vastly improved the efficiency with which they employ their indigenous technical resources. Although European and Asian productivity growth rates have long exceeded that of the United States, by the late-1980s the most advanced of these countries had finally closed the gap with the United States in absolute manufacturing productivity (Figure 1.3).

Granted, a comparison of overall productivity rates (Figure 1.4) shows that the United States continues to enjoy an absolute advantage over its major competitors. Considering the relatively poor U.S. performance in manufacturing productivity growth over the past two decades and the fact that manufacturing accounts for less than a quarter of the combined output of the Organization for Economic Cooperation and Development (OECD) countries (only 20 percent of U.S. output), these figures attest to the high productivity of the U.S. nonmanufacturing sectors relative to their counterparts in Western Europe or Asia.[5] Perhaps reflecting the singleness of purpose with which Japan has developed its export-oriented manufacturing industries, the dismal productivity of Japan's nonmanufacturing and non-tradable sectors has dragged the nation's overall output per person employed to the lowest level of The Group of Seven (G-7) economies—Canada, Federal Republic of Germany, France, Italy, Japan, United Kingdom, and the United States.

As in the case of inputs into a nation's technological enterprise, there are any number of ways that the technical output of a country can be measured, each with its own special insights and limitations. Patent data, for example, tell little about a country's or a firm's ability to commercialize its innovations. Yet, for many industries, patent data provide a useful window on the pure technical strength of nations or firms.[6] Between 1978 and 1988, the share of total patents granted in the United States to U.S. inventors fell from 62.4 to 52 percent. The U.S. decline was directly offset by a doubling of the Japanese share from 10.5 to 20.7 percent, while the share of European inventors remained unchanged at around 18 percent (Figure 1.5). Over the period, relative Japanese patent performance in high-tech[7] products such as computers, communications equipment, and electronic components was particularly impressive (Figure 1.6). The only high-tech product field in which the share of patents to U.S. inventors increased over the period was "drugs and medicines."

Recent changes in national shares of world production, trade, and foreign direct investment in high-tech industries confirm the shift in the technical balance of power suggested by patent data. Between 1975 and 1986, world production of high tech manufactures experienced a sixfold increase and world high-tech trade underwent a ninefold expansion (Figures 1.7 and 1.8). Over the same period, Japan nearly doubled its share of both world production and exports of high-tech products, displacing the United States as the world's leading high-tech exporter in the process.[8]

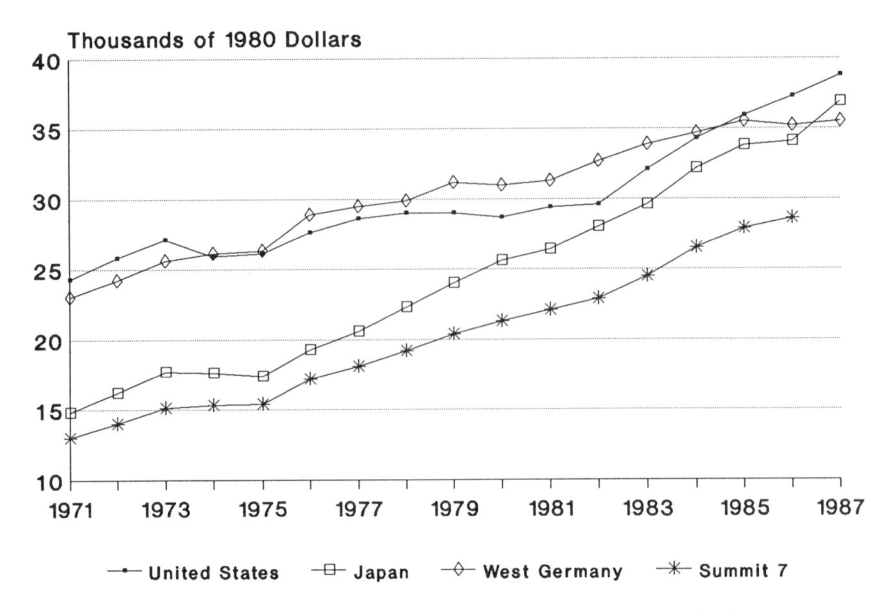

FIGURE 1.3 Manufacturing output per manufacturing employee, trends in absolute growth: 1971–1987, in constant 1980 dollars. Average for Summit 7 includes France, Italy, Japan, and United Kingdom. SOURCE: Council on Competitiveness (1990).

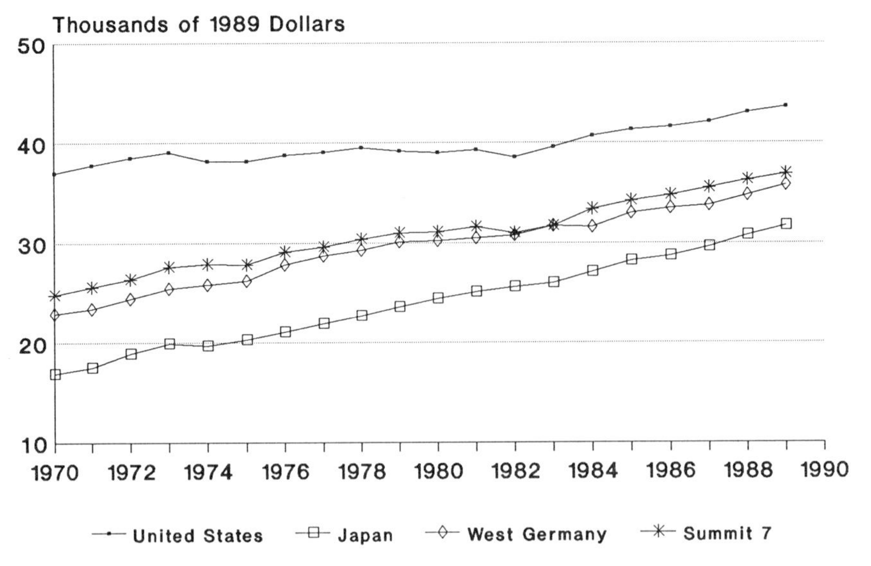

FIGURE 1.4 Gross domestic product per employed person, 1970–1989, purchasing power parity exchange rates. Average for Summit 7 includes Canada, France, Italy, and United Kingdom. SOURCE: U.S. Department of Labor (1990).

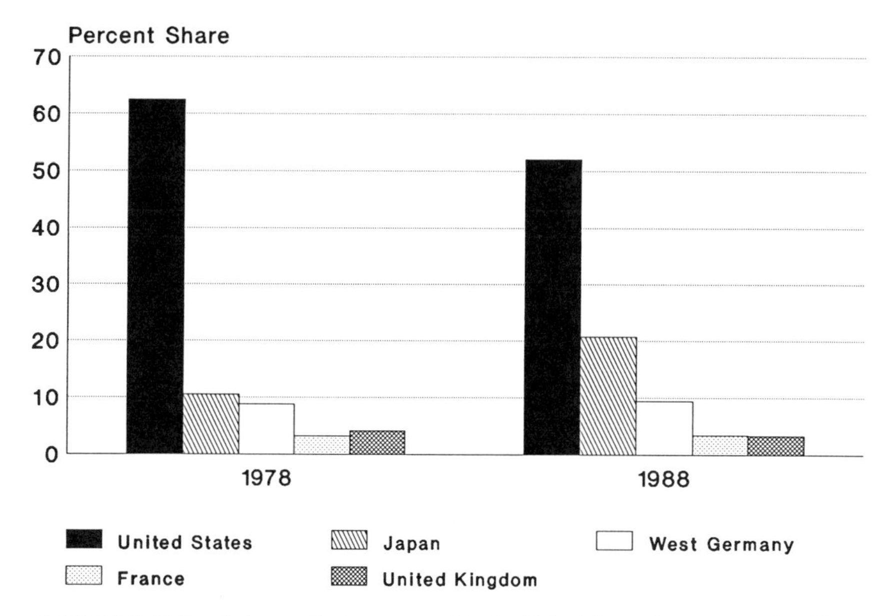

FIGURE 1.5 National shares of patents granted in the United States, by country of residence of inventor and year of grant, all technologies: 1978 and 1988. SOURCE: National Science Board (1989, p. 362).

Although the United States continues to produce a larger volume of high-technology products than any other nation, its share of world high-tech output (42 percent) remained relatively stable during the 1970s and 1980s while that of Japan grew dramatically from 18 percent in 1975 to 32 percent in 1986. Over the same period, European nations watched their share of world high-tech output drop from 36 to 24 percent.

The sharp expansion of European and Asian outward foreign direct investment during the past two decades offers a striking expression of the enhanced technological competence and confidence of foreign corporations. Since 1973 there has been a fivefold increase in the volume of world foreign direct investment and a significant redistribution in shares of total outward foreign direct investment among the major industrialized countries (Figure 1.9).

Between 1973 and 1987, the U.S. share of world outward foreign direct investment declined from 48 to 31.5 percent, while that of the Western European countries expanded from 39 to 51.2 percent and Japan's share rose from 0.7 to 7.5 percent. From 1975 to 1985, the stock of foreign direct investment in manufacturing accounted for by the G-5 economies doubled while the U.S. share of that total declined from 58 to 46 percent. Meanwhile, the share of the combined foreign direct investment stock in manufacturing held by European corporations jumped from 35 to 38 percent

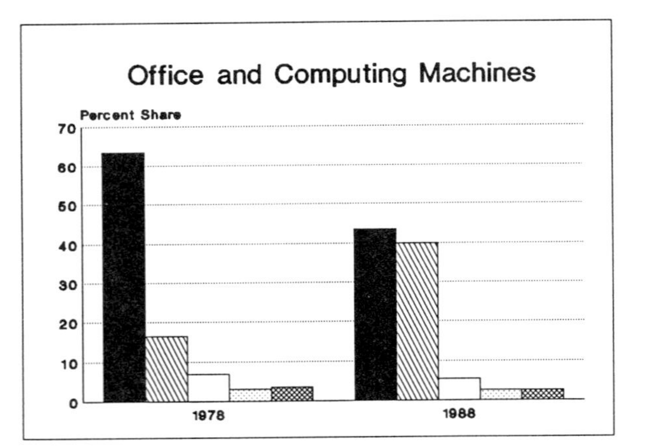
Office and Computing Machines
Percent Share
70
60
50
40
30
20
10
0
1978
1988

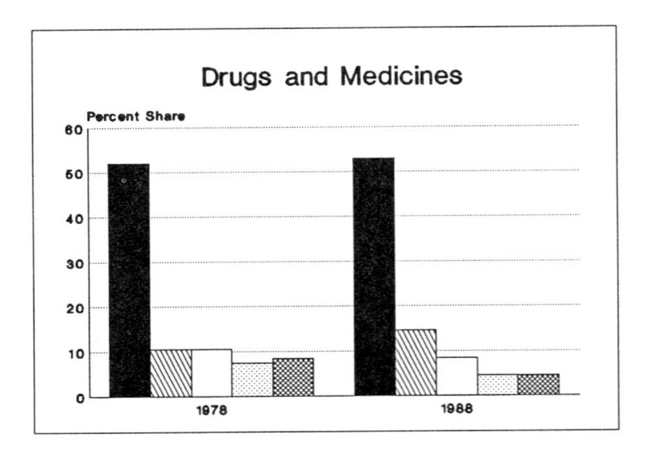
Drugs and Medicines
Percent Share
60
50
40
30
20
10
0
1978
1988

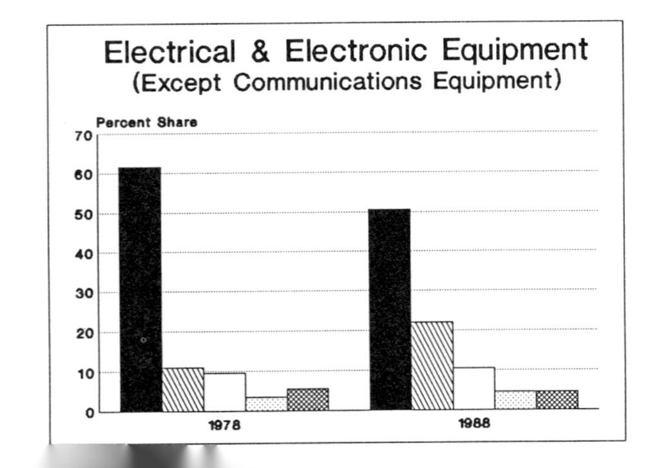
Electrical & Electronic Equipment
(Except Communications Equipment)
Percent Share
70
60
50
40
30
20
10
0
1978
1988

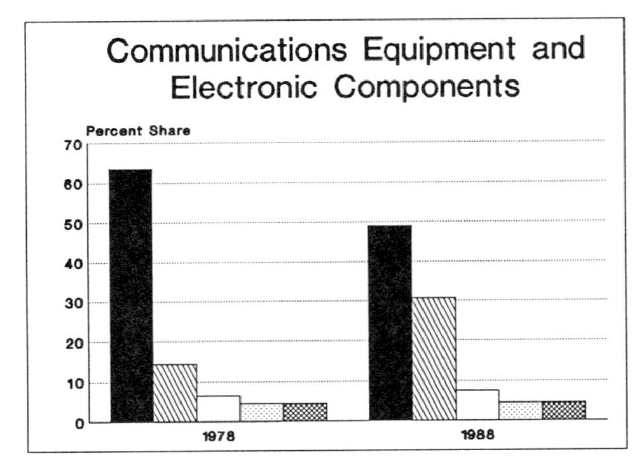
Communications Equipment and
Electronic Components
Percent Share
70
60
50
40
30
20
10
0
1978
1988

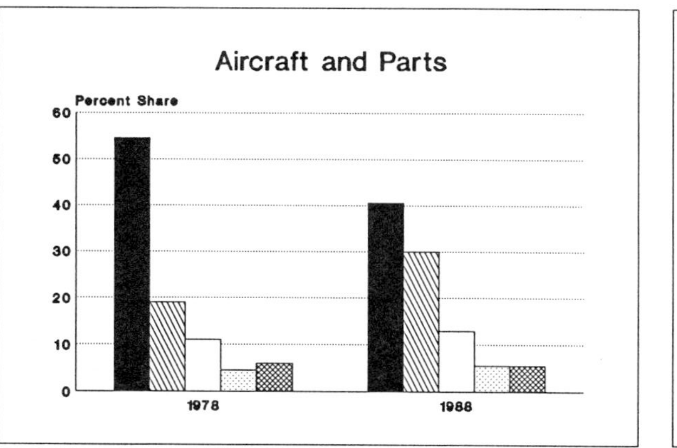

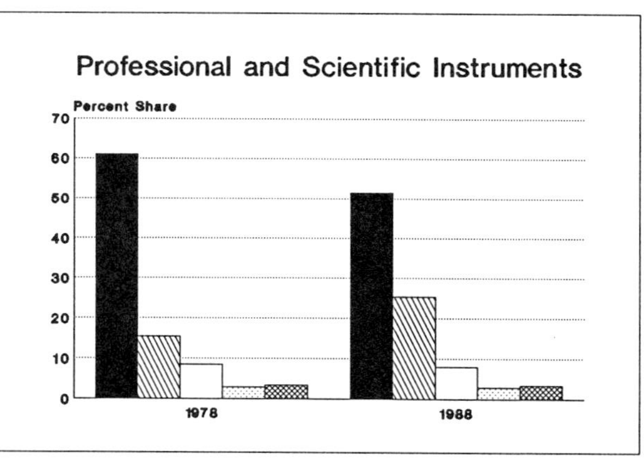

FIGURE 1.6 National shares of patents granted in the United States, by country, product field, and year of grant: 1978 and 1988. SOURCE: National Science Board (1989, p. 362).

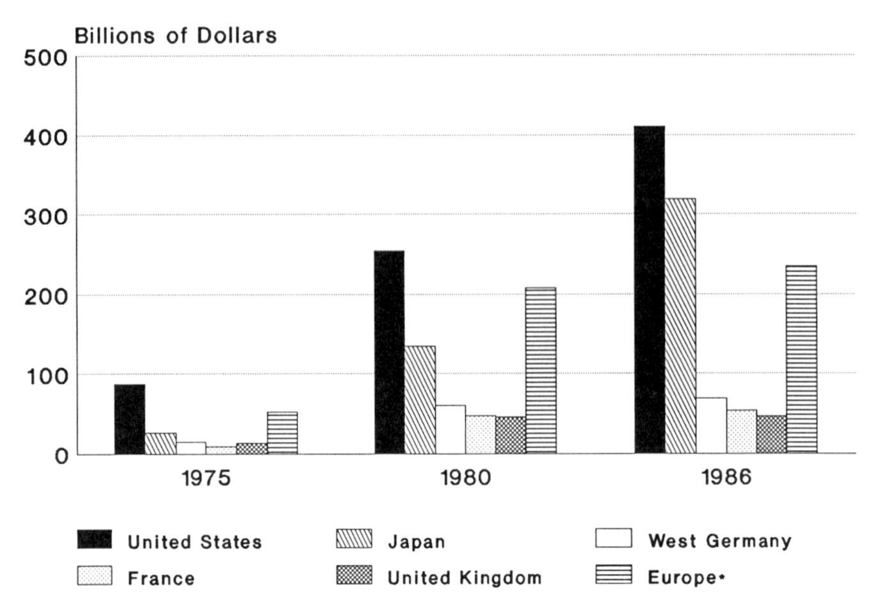

FIGURE 1.7 Global production of high-technology products, by selected countries: 1975, 1980, and 1986. SOURCE: National Science Board (1989, p. 371).

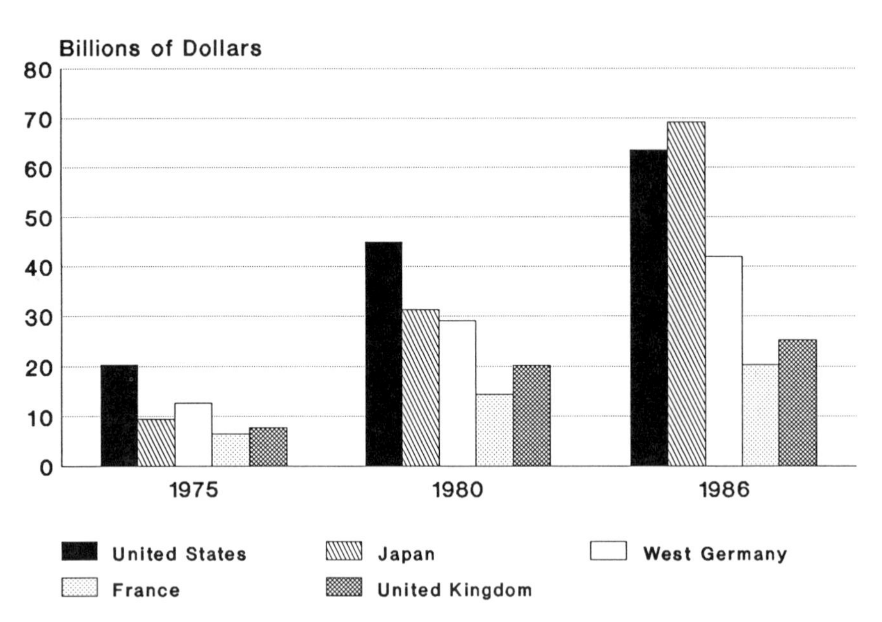

FIGURE 1.8 Exports of high-technology products, by selected countries: 1975, 1980, and 1986. SOURCE: National Science Board (1989, p. 377).

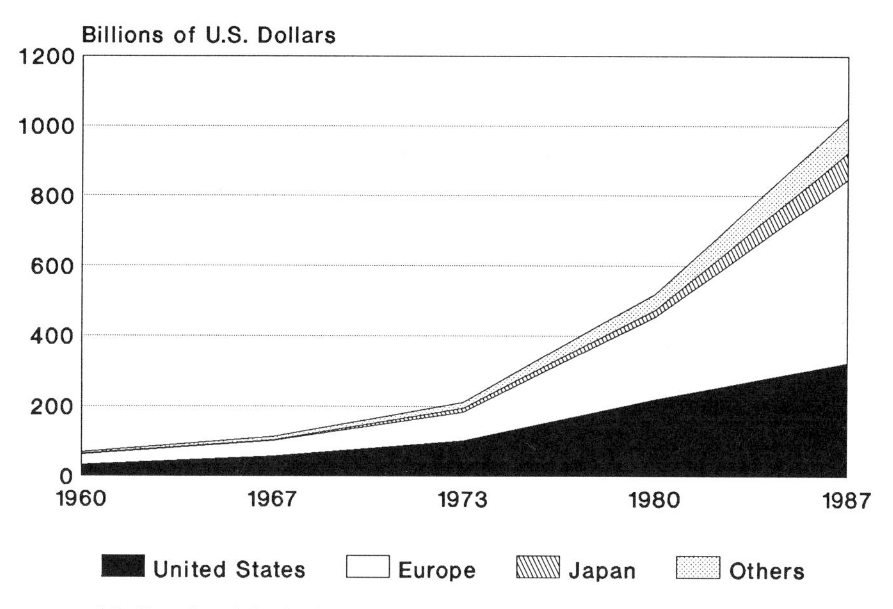

FIGURE 1.9 Growth and distribution of world outward stock of foreign direct investment by country of origin: 1960–1987. SOURCE: U.S. Department of Commerce (1989b, p. 11).

and that of Japanese corporations from 7 to 15 percent (United Nations Centre on Transnational Corporations, 1988; U.S. Department of Commerce, 1988c, 1989b).

In summary, there has been a dramatic shift during the past two decades from a technologically unipolar world, led by the United States, to one in which technological capabilities are much more dispersed among a number of industrialized and industrializing countries. This sea change in the global technological order and the accompanying intensification of international competition have had profound implications for the organization of corporate technical activities across national borders.

INTEGRATION OF NATIONAL TECHNOLOGY ENTERPRISES SINCE THE MID-1970s

The integration of national technology markets has been gathering momentum since the early 1950s, fueled largely by the postwar expansion of world trade and the growth of predominantly U.S. multinational business activities. Yet, until recently, the pace and scope of global technical integration have been significantly circumscribed by the highly uneven distribution of technical capabilities worldwide. To be sure, international transfers of commercial and military technology were significant during the 1950s and 1960s. However, the unchallenged technological and industrial supremacy of the United States guaranteed that technology flows were predominantly "one-way," that is, from the United States to the rest of the world. This situation, in turn, tended to ensure that the advanced technical activities associated with the research, design, and development of products or production processes for most industries remained organized more along national than multinational or global lines.

Before the 1970s, U.S. and foreign multinational corporations in manufacturing industries tended to develop and commercialize most new products and technologies within their home markets first, transferring production abroad only after product and process technologies were more mature or standardized.[9] In other words, the most sophisticated, most proprietary, or most highly leverageable technical activities (research, product and process development, design, systems integration) were generally concentrated in the home market while the less sophisticated, more standardized technical functions (manufacturing, assembly, component and capital equipment production) were often transferred to subsidiaries overseas.[10] In short, the technology base of most industries remained essentially national even as production became increasingly multinational.

During the last decade and a half, however, there has been a fundamental shift in the international organization of production and advanced technical activities. Unlike the internationalization of production during the 1950s and 1960s, which was driven primarily by U.S. foreign direct investment, internationalization since the mid-1970s has been characterized by a rapid expansion of non-U.S. foreign direct investment and a proliferation of transnational corporate alliances. In the last decade alone, world foreign direct investment has doubled, growing four times as fast as world trade since 1983 (Figure 1.10). By 1987, however, the U.S. share of world outward foreign direct investment had declined to 31.5 percent, down nearly 17 percentage points from its share in 1973 of 48 percent (see Figure 1.9). Since 1980 there has also been a rapid increase in the formation of transnational corporate alliances, most of these initiated by U.S. firms (see Hagedoorn and Schakenraad, 1990a,b).

These two new trends in the internationalization of production combined with the intensification of international competition, the cross-penetration of national markets, and the rapid spread of advances in information and production technologies, have propelled the world's largest, and, historically, most self-sufficient national economy to unprecedented levels of economic and technical interdependence. Moreover, they have brought about the transnationalization of the technology development and acquisition strategies of corporations in a growing number of industries.

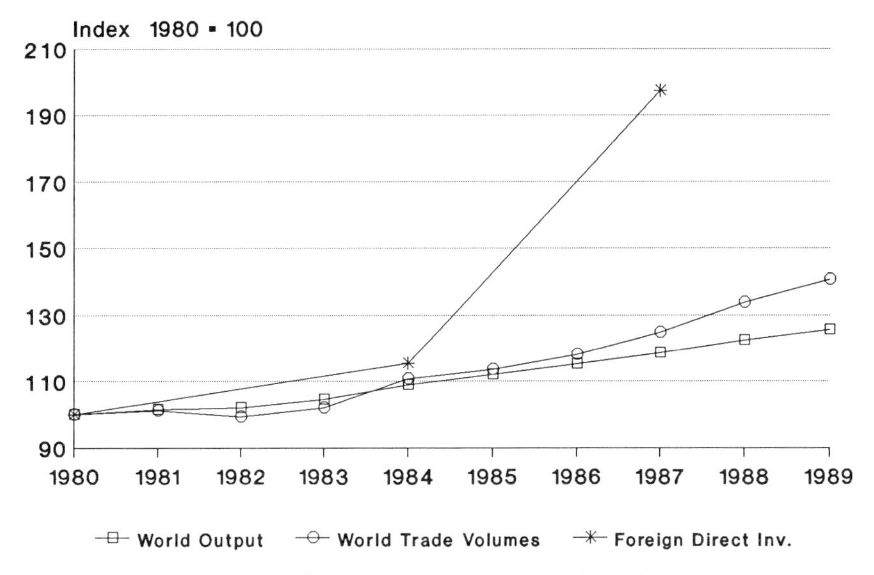

FIGURE 1.10 Growth of world trade, output, and foreign direct investment: 1980–1989. SOURCES: International Monetary Fund (1990, p. 73 and p. 93), U.S. Department of Commerce (1989b, p.11), and U.S. Department of Commerce (1988c, p. 87).

GROWTH OF U.S. ECONOMIC AND TECHNOLOGICAL INTERDEPENDENCE

The increase in U.S. economic and technological interdependence since the late 1970s suggests how rapid and pervasive the integration of national technical enterprises has been during the past decade. Between 1980 and 1986 alone, U.S. imports of foreign high-tech products increased from 11.5 to 18.1 percent of total domestic consumption. In 1986 the United States experienced its first high-technology trade deficit since data collection began. Between 1973 and 1987, as the volume of world foreign direct investment experienced a fivefold increase, the U.S. share of total inward foreign direct investment grew from 9.9 to 25.2 percent. Moreover, between 1977 and 1987, the stock of foreign direct investment in U.S. manufacturing grew from about 5 percent to more than 12 percent of total U.S. manufacturing assets. Currently, it is estimated that foreign-owned firms operating in the United States account for nearly a quarter of U.S. exports and one-third of U.S. imports (Graham and Krugman, 1989; Julius, 1990; National Science Board, 1989; U.S. Department of Commerce, 1989b).

The dependence of U.S.-owned high-technology firms on overseas markets and the productive capabilities of foreign affiliates has also grown in recent years. In 1986, sales of high-technology products by the foreign affiliates of U.S. companies were twice as large as U.S. high-technology exports, and the ratio of foreign affiliate to U.S. parent assets for high-technology manufacturing industries stood at nearly 42 percent. Moreover, recent studies suggest that the reverse transfer of technology from the foreign affiliates of U.S. companies to their U.S.-based parents has grown substantially in volume and importance over the course of the past two decades (Mansfield and Romeo, 1984; National Science Board, 1989).[11]

The growing dependence of the U.S. economy on foreign technical talent during the 1970s and 1980s is equally remarkable. In 1972 foreign-born engineers represented less than 8 percent of the total U.S. engineering work force. By 1982, however, their share had risen to nearly 18 percent, a large proportion of which were engaged in industrial and academic research and development. Similarly, between 1975 and 1985, the share of all U.S. engineering faculty members under the age of 36 accounted for by foreigners increased from 10 percent to nearly 50 percent (National Research Council, 1988).

CHANGING CORPORATE STRATEGIES TOWARD TECHNOLOGY DEVELOPMENT AND ACQUISITION

The rapid growth of international trade and direct investment also signaled a shift in corporate strategies with profound implications for the organization of technical activities across national borders. The intensity of global competition unleashed during the late 1970s forced national and transnational corporations in a number of industries to recast their strategies and restructure their operations in more global terms. The clearest expression of this change in corporate strategies has been the acceleration of foreign direct investment itself as European, Asian, and American companies have moved to penetrate each other's home markets. Another element of this corporate response, perhaps most evident in the strategies of U.S. corporations, has been a trend toward increased internationalization, decentralization, and foreign development and acquisition of a growing range of advanced technical activities, including research, design, and development. Although much of this shift in corporate strategy has been accomplished through foreign direct investment and trade, another important instrument for implementing new strategies has been transnational technical alliances among corporations.

The reasons for these changes in the way more and more national and transnational companies are organizing and conducting their advanced technical activities are multiple and vary considerably in relative importance from one industry to the next. Nevertheless, some of the most frequently cited explanations that seem to cut across industries include the desire of firms to access new markets, to better monitor the capabilities of competitors and their home markets, to exploit unique technical, managerial, and other operational capabilities of would-be potential partners, to accelerate the process of innovation and reduce product and process development time, or to share the risks and costs involved in research, development, and production. Evidence of this trend, though mostly based on company or industry case studies, finds at least some quantitative expression in recent surveys of U.S. corporate R&D spending and interfirm alliances (Chenais, 1988; Enderwick, 1989; Hagedoorn and Schakenraad, 1990a,b; Mowery, 1988a; National Science Foundation, 1989a; Vonortas, 1989). In aggregate terms, overseas R&D spending by U.S. corporations nearly doubled between 1979 and 1987, yet foreign R&D spending as a share of total R&D spending by U.S.-owned corporations actually declined over the period from 10.7 to 7.7 percent. Nonetheless, leading U.S. companies in the computer, telecommunications, microelectronics, pharmaceuticals, and automotive industries are reported to conduct anywhere from a quarter to a third of their R&D activities abroad.[12]

Conversely, there is evidence that the surge of foreign direct investment into the United States during the past decade has brought with it significant expansion of U.S.-based R&D activities by foreign transnational corporations. In 1982, in-house R&D spending by the U.S. affiliates of foreign-owned firms surpassed that of U.S. corporations abroad for the first time. Four years later, U.S. affiliates of foreign companies spent more than $5.5 billion on in-house R&D, nearly a billion dollars more than the foreign affiliates of U.S. companies (National Science Foundation, 1989a, U.S. Department of Commerce, 1984–1988b).[13] Lack of data prevents useful comparisons of the relative importance of overseas R&D activity for individual foreign countries, industries, or firms. However, based on committee discussions and interviews with foreign corporate representatives it is estimated that the advanced technical activities of U.S. transnational corporations are generally more internationalized than those of their European or Asian counterparts. This makes intuitive sense, given that most Asian corporations and many European corporations are relative newcomers to the world of transnational production.

By far the most obvious manifestation of recent changes in the R&D and technology sourcing strategies of corporations has been the proliferation of private transnational technical alliances and networks since the late 1970s (Figure 1.11). The causes of this sudden upsurge of alliance activity are manifold and complex. The emergence of a more plural global technical order with multiple, large technically advanced national and regional markets, the intensification of global competition, and the shortening of product cycles have all played important roles in this shift in corporate strategy. Moreover, policy-induced barriers to trade and investment appear to have been equally determinant in compelling or encouraging globally active cor-

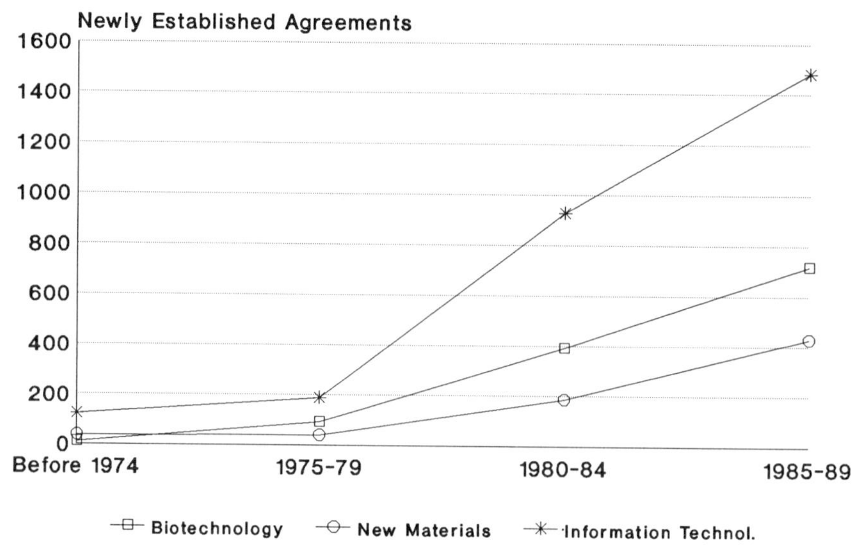

FIGURE 1.11 Growth of newly established technology cooperation agreements in biotechnology, information technologies, and new materials: 1974–1989 (1989 first seven months only). SOURCE: Hagedoorn and Schakenraad (1990a, p. 6, p. 7, p. 9).

porations to ally themselves with other firms in important foreign markets (Mowery, 1988a). The importance of these "environmental" changes is reflected in actual surveys of corporate alliance participants. Hagedoorn and Schakenraad (1990b), for instance, found that four motives appeared to play a particularly significant role in the establishment of interfirm technical alliances: (1) the search for and access to new markets, (2) the technological complementarity of prospective partner, (3) the reduction of the innovation time-span, and (4) a desire to monitor technological opportunities.

As in the case of overseas R&D spending, U.S. companies appear to have taken the lead in building transnational technical alliances. According to recent studies by Hagedoorn and Schakenraad (1990a,b), U.S. firms were involved in nearly 70 percent of all newly established technical alliances in biotechnology, about 60 percent of those in information technology, and more than 50 percent of alliances in new materials between 1970 and 1989. More than half of all agreements involving U.S. firms in all three technology areas were transnational in scope (U.S.-Western Europe or U.S.-Japan).[14]

Interfirm technical networks in all three technology areas have become increasingly dense over the course of the 1980s, that is, both the number of alliances between previously allied firms and the number firms interconnected through major nodes in each technical network grew dramatically over the period.[15] Furthermore, the networks formed by interfirm alliances are becoming increasingly "international" over time, that is, the pattern of national or regional blocs of interfirm alliances, most pronounced in the biotechnology and materials fields to begin with, has begun to give way to greater interbloc technical collaboration. The "internationalization" of interfirm technical networks is particularly apparent in subsets of information technology such as microelectronics.

Debate continues as to whether the rapid growth of transnational technical alliances in the 1980s will enhance or impede global technological advance and competition in the future. There are even considerable differences of opinion as to whether the proliferation of corporate alliances constitutes a long-term reorientation of corporate strategies or a more ephemeral phenomenon. One thing is clear, however. The alliance boom of the past 10 years has contributed significantly to the emergence of a truly transnational technology base for a range of industries.

INTERINDUSTRY VARIATIONS IN THE SCOPE AND CHARACTER OF GLOBALIZATION

In an effort to explore interindustry variations in the degree to which technical activities have become transnational in scope and to assess the common implications of globalization for U.S. industry as a whole, the committee evaluated the changing character of global technological competition in eight technology-intensive industries.[16] Corresponding to the committee's areas of expertise, the case studies included technically mature industries such as automotive, construction, electrical apparatus, and petrochemicals, as well as technologically more dynamic industries such as aircraft engines, computer printers, semiconductors, and the emerging industry of biotechnology.[17] Examining each industry's value-added chain, from materials, components, and capital equipment through manufacturing and assembly to distribution, sales, and service, the committee attempted to identify a short list of industry-specific "critical" technologies and to evaluate the relative technical strengths and weaknesses of U.S.-based producers in each industry.[18]

Comparison of the recent globalization of technical activities in three industries studied by the committee—automotive, aircraft engine, and construction—illustrates the diversity of industry experience, while pointing to a number of cross-industry commonalities.

Automotive Industry

Despite the fact that there have been multinational auto companies since the industry's commercial takeoff in the early 1900s, the automotive industry has remained essentially a national industry in terms of the organization and conduct of most technical and commercial activities throughout the industry's value-added chain. For most of the postwar period, foreign direct investment has been the predominant mode of gaining access to foreign markets, trade in motor vehicles being only of secondary importance. The high-volume U.S. automakers dominated multinational activity in the industry as they sought to jump trade barriers, access low-cost production locations, and leverage their special technical, financial, and organizational capabilities across a wide range of national markets. Primarily because of the activities of multinational companies in both the vehicle assembly and auto supply sectors, the industry's basic technology has been relatively homogeneous and widely diffused internationally for decades.

Beginning in the mid-1970s, growing evidence of worldwide excess capacity combined with a powerful surge in the competitiveness of the Japanese auto industry to provoke a major transformation in the global strategies and conduct of the world's leading automakers. Led by the rapid expansion of Japanese foreign direct investment and the proliferation of transnational alliances and joint ventures by both automakers and their suppliers, the value-added chain of the automotive industry, including more and more advanced technical activities, has become increasingly transnational in organization.

The motives behind the surge in Japanese foreign direct investment were essentially the same as those that inspired U.S. automakers to invest overseas in the past—to improve market access, that is, jump trade barriers, and leverage their special manufacturing capabilities in a large market. More novel was the expansion of transnational corporate alliances and joint ventures including codesign, coproduction, cosourcing, and joint distribution agreements. These interfirm arrangements complemented the Japanese foreign direct investment drive by offering Japanese carmakers the opportunity to defuse protectionist sentiment within their host market, to share costs and reduce risks of expanding their presence in the host market, and, perhaps most important, the chance to move up the learning curve regarding the peculiarities of their host market much more rapidly with the help of local partners. On the other hand, interfirm agreements offered U.S. and European automakers relatively inexpensive mechanisms for tapping the special technical and organizational know-how of their Japanese counterparts (Womack, 1988).

In this new global engineering and production environment, there is a growing consensus among U.S. automakers that competing successfully at home and abroad depends increasingly on improving their ability to manage the entire design-development-manufacturing process as efficiently as their Japanese competitors. In terms of pure technological capabilities, the big three U.S. automakers remain among the industry's vanguard, and their extensive global distribution and marketing systems are still formidable competitive assets. Nevertheless, despite (or perhaps because of) their early leadership in the development and application of mass production technologies, U.S. automakers have lagged behind their Japanese competitors in the application of new automated flexible manufacturing technologies and the overall management of the product development cycle. Furthermore, the historical pattern of relations between the U.S. assembly industry and its supplier base (both captive and merchant) demonstrates a long-standing disregard by assemblers and suppliers for the additional organizational learning and technological advance that their European and Japanese counterparts have achieved through closer collaboration up and down the industry's value-added chain. The U.S. automotive industry is beginning to redress some of these liabilities, most notably through international technical and commercial alliances. Nonetheless, the U.S. industry's global strategy continues to focus on ways to better leverage or manage its global resources and "rationalize" global production, and thereby meet competition at home and abroad more effectively (see Appendix A).

Construction Industry

For the most part, construction is inherently a "national" or "local" industry. Given the highly local character of construction labor markets, building codes, and building materials markets, most construction activity is protected or precluded from international competition. The only segments of the industry that have experienced any significant degree of internationalization are nonresidential construction and design engineering services. The predominant modes of international transaction in these more specialized construction services have traditionally been through "long-distance" trade (e.g., preparing blueprints to be used by local contractors overseas) or services rendered on-site in a foreign country through branch operations of an international construction firm. Until the late-1970s, most international trade in construction services was conducted between a relatively small number of international firms located in industrialized countries and governments or private parties in developing or industrializing countries. The industry's technology base, on the other hand, has long been globalized, largely as a result of the highly internationalized nature of the construction equipment industry and international cooperation in construction research.

Since the early 1980s, however, trade in nonresidential construction and design services among the industrialized nations of Japan, North America, and Western Europe has grown more rapidly than that between developing and developed countries. This shifting pattern of trade has been accompanied by several changes in the character of internationalization itself. First, international mergers and acquisitions have become an increasingly important vehicle for internationalization. This phenomenon is most pronounced in the relatively open construction markets of North America and Western European, while virtually nonexistent in the more closed Japanese market. Second, there has been an increase in reliance by construction and design engineering companies on international joint ventures to work domestic and foreign markets. Some industry analysts interpret this development as a harbinger of greater international specialization within the industry. The primary motives for these changes in the strategies of international construction and engineering design firms were generally to improve market access, to acquire or otherwise join forces with the special technical and managerial capabilities of other firms, and to share costs and risks.

Finally, there has been an increase in international cooperation in construction research and development. Participation by international construction firms in transnational R&D is highly uneven because the locus of R&D activity in the United States and Western Europe is outside the firm, that is, in government, university, supplier, or trade association laboratories. In Japan, however, R&D is primarily the responsibility of the construction companies proper. In part because of this structural difference, Japanese firms have been much more aggressive than their U.S. or European counterparts about establishing links with construction research abroad.

A comparison of the U.S. international construction industry with its counterparts in Western Europe and Japan highlights the U.S. industry's relative engineering strengths and weaknesses. Like their American competitors, European international construction firms have established reputations as project managers, albeit with different areas of expertise. U.S. firms excel in the design and construction management of large, complex projects, such as airports, petroleum refineries, and nuclear power plants. European firms are most competitive in the area of smaller, more specialized projects. Both European and U.S. firms procure equipment and materials worldwide, and their home markets are more or less open to foreign penetration (government procurement requirements notwithstanding). However, European construction firms enjoy closer working relationships with their equipment providers than their U.S. competitors do.

The Japanese industry, by contrast, is much more vertically integrated and, as a result, much more parochial in its sourcing of material and equipment. Japanese international construction firms also benefit from close links with Japanese real estate developers and with Japanese manufacturing firms (and their foreign subsidiaries) in overseas markets. As noted, the Japanese are also far more involved in research and development of advanced construction methods and equipment than their U.S. or European competitors. Their close ties with Japanese international real estate development notwithstanding, Japanese international construction firms compete primarily on the basis of their technical competence and reliability. Unlike

European and American firms, Japanese construction companies have relatively little reputation internationally in project management (see Appendix A).

Aircraft Engine Industry

The production and technology base of the aircraft engine industry has been organized along national lines for most of the industry's history, even though trade in aircraft engines has long been significant. The reasons for the historically "national" orientation of the industry's production and technical activities have much to do with the nature of its product, production processes, and markets. First, the sheer technical complexity of this industry's product fostered concentration within the aircraft industry, raised monumental barriers to market entry for any latecomers, and has tended to discourage foreign direct investment. Moreover, this complexity was coupled with a corresponding demand for expensive, highly sophisticated human and physical capital, and heavy R&D requirements that span the entire product life cycle (approximately 30 years). Second, the rapid growth of U.S. domestic demand for aircraft engines and extensive federal support for R&D, testing and work force development offered few incentives for the leading U.S. firms to internationalize their advanced technical activities. Finally, the importance of the industry to national defense caused governments to circumscribe the internationalization of aircraft engine production and technology development.

Beginning in the late 1960s, a number of changes in domestic and international markets led the technically dominant U.S. firms, General Electric and Pratt & Whitney, to enter into a wide range of technical alliances, joint ventures, and component sourcing agreements with the only other full-range engine manufacturer, Rolls Royce, and a dozen or so second-tier engine manufacturers in North America, Europe, and Japan. First, the cost of developing and launching new models of engines for commercial use began an inflationary spiral that would continue into the 1990s. Second, during the 1970s and 1980s, there was a general decline in the level of federal support for civilian research funding and manpower through the National Aeronautics and Space Administration (NASA). These two developments have effectively shifted a greater share of the cost and risk of technological advance from the federal government to the engine manufacturers themselves. Finally, the 1970s witnessed a decline in the relative size of the U.S. market for commercial aircraft and growth in the relative size of foreign markets (Mowery, 1988b).

This combination of spiraling development costs and the growth of foreign markets provided the impetus for a wave of transnational alliances between the first- and the second-tier companies, and among second-tier companies themselves, to increase market access, share technology, and reduce fixed costs (sharing risk). In the process, this shift in corporate strategies has greatly increasing the scope of international technological and commercial interdependence in the industry.

The pattern of comparative national specialization is less pronounced in the aircraft engine industry, largely because U.S. firms have held such a commanding lead in the industry's technology for so many decades. After all, the fact remains that no new primary national manufacturer of commercial engines for mainline jet transports has emerged for 25 years—General Electric, Pratt & Whitney, and Rolls Royce have it all. Nonetheless, developments of the past two decades have forced the two leading U.S. engine manufacturers to increase their dependence on out- or foreign-sourced components, materials, and manufacturing capabilities. This, in turn, has led them to increase their focus on research, design, and system's integration. Despite the continuing lead of U.S. companies in most of the industry's critical technologies, such as aerothermodynamics and structural design, European and Japanese competitors have demonstrated competitive advantages in the application of advanced manufacturing processes, and various aspects of materials and controls. Moreover, as a result of the sustained European effort in the commercial aircraft industry, European engine manufacturers are also closing the gap with the United States in structural design and systems integration (see Appendix A).

Cross-Industry Commonalities

A comparison of the globalization experiences of the automotive, construction, and aircraft engine industries, as well as those of the other industries surveyed by the committee (see Appendix A), underlines a number of commonalities. First, in all of the industries studied the technical capabilities of the three major industrialized regions, North America, Western Europe, and Japan, appear to have undergone significant convergence since the early 1970s. Second, this redistribution of technical strength has been accompanied by a growing cross-penetration and integration of the national technology base for each industry by way of transnational alliances, foreign direct investment, or the expansion of international trade. Third, in almost all of the industries studied, U.S.-owned transnational corporations appear to have taken the lead in globalizing the industry's technology base, either by developing or acquiring a greater share and range of their advanced technical activities abroad, or by trading technology and know-how for market access more aggressively than their foreign competitors.

On the other hand, comparisons of the relative performance of U.S. producers with respect to particular "critical" technical and managerial functions across industries suggest a common pattern of U.S. technical strengths and vulnerabilities. In virtually every industry studied by the committee, U.S. producers appear to have lost the most ground to foreign competition in the development, application, adaptation, and management of advanced process technology related most closely to manufacturing proper, whether of final goods, subassemblies, components, or capital equipment.

This handicap has been particularly pronounced in U.S.-based industries in which relationships between firms within an industry's value-added chain (suppliers, assemblers/systems integrators, and customers/users) have been intensively "arms length," such as in the construction, automotive, or semiconductor industries. However, it is also acknowledged as a persistent competitive vulnerability in more vertically integrated or "networked" U.S. businesses, such as the aircraft engine and computer printer industries.

Alternatively, U.S.-based companies appear to have retained leadership in the more prestigious technical areas of product design and development and the integration of complex systems. This is particularly apparent in industries where (a) U.S.-based companies have effectively managed and controlled integration of the system of production and distribution either through vertical integration or effective use of interfirm relationships (for example, with their supplier base, technology partners, or licensees), or (b) the product or process of production depends on highly sophisticated applications software or rapidly changing science and can be executed by small or growing companies (advanced materials, biotechnology, etc.).

GLOBALIZATION OF U.S. UNIVERSITY-BASED TECHNICAL CAPABILITIES

Along with U.S. multinational corporations, U.S. universities have long been a primary driver of the globalization of technology. Through education of foreign students, the employment of foreign faculty and research associates, and a firm commitment to the free flow of knowledge without regard to national borders, U.S. university science and engineering departments have played a central role in international technology transfer. Between 1955 and 1985, the number of foreign students studying engineering and science at U.S. universities increased by a factor of 10, and more than half of these obtained graduate degrees from their host institution. Over the same period, the flow of foreign postdoctoral researchers and visiting faculty through U.S. research universities has experienced similar growth.

For most of the period since World War II, the relationship between the U.S. university-based technical enterprise and its foreign clients and counterparts has been characterized by lopsided dependence of the latter on the U.S. academic "mecca." However, as with the U.S. industry-based technical enterprise, U.S. universities have watched one-sided international dependence give way to complex interdependence over the past decade and a half.

The most dramatic expression of this growing interdependence is provided by changes in the ratio of foreign to domestic graduate students and faculty in U.S. engineering schools since the mid-1970s. Undergraduate engineering education has remained a "national" enterprise in which foreign students have represented less than 10 percent of total enrollment since data collection began in the 1950s. However, in 1975, the share of foreign-born graduate students and faculty in U.S. engineering schools, which had been relatively stable since the mid-1950s, began a decade of unprecedented expansion. By 1985, foreign-born students accounted for 50 percent of engineering doctoral candidates and nearly two-thirds of all engineering postdoctoral researchers at U.S. universities. In 1975, only 10 percent of U.S. engineering faculty members under the age of 36 were foreign-born. Ten years later the foreign share stood at 50 percent (National Research Council, 1988).

The sudden rise in the foreign-born shares of total graduate enrollment, postdoctorates, and faculty employment is a function of three interrelated developments: (1) the rapid growth of university research activities during the past decade, and with it, a rapid increase in demand for research personnel; (2) an equally rapid increase in the demand from U.S. industry for engineering graduates (mostly B.S. recipients); and (3) a prolonged slump in the number of U.S.-born engineers and engineering students deciding to pursue doctoral degrees in engineering or an academic career during the 1970s and early 1980s.

Between 1978 and 1988, the U.S. academic research budget for engineering disciplines doubled in real terms from roughly $1 billion to $2 billion. Over the same period, total graduate enrollment in U.S. engineering programs grew at an average annual rate of nearly 6 percent. Paralleling the rapid expansion of the university research enterprise, a prolonged upswing in demand by U.S. industry for engineering graduates, mostly B.S. engineers, indirectly fueled university demand for engineering faculty. From 1972 to 1986, total engineering employment growth in the United States averaged 7 percent per year. More than 80 percent of that growth was accounted for by B.S. engineers. Yet, for most of the past 20 years, while demand for engineering graduate students and faculty was increasing, the absolute number of U.S.-born engineers and engineering students deciding to take Ph.D. degrees in engineering or to enter the teaching profession declined[19] (see Figure 1.12).

Unable to attract an adequate supply of U.S.-born engineering bachelor degree holders, U.S. university engineering doctoral programs have been forced to look abroad for students to keep their research programs fully engaged.[20] Because faculty are drawn from the population of academically oriented new Ph.D.'s, the trends in graduate student enrollment produce similar trends in faculty composition.

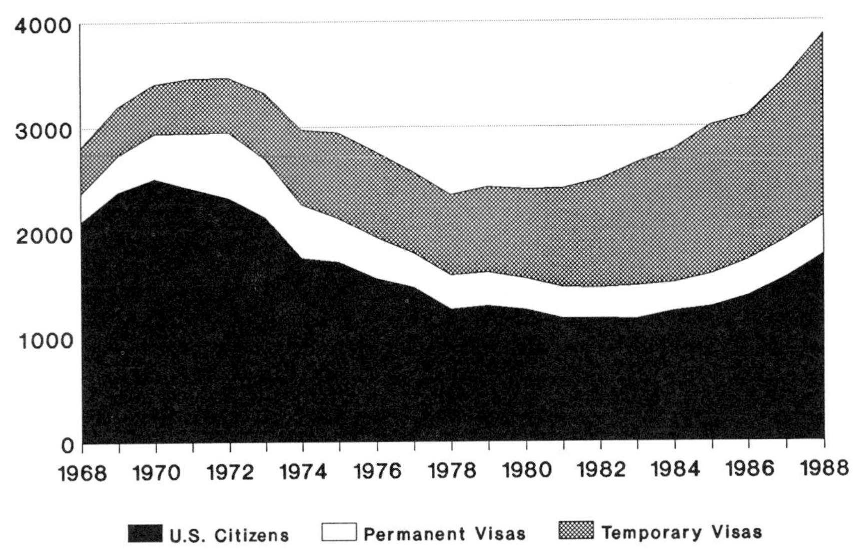

FIGURE 1.12 Engineering Ph.D. awards in the United States, by citizenship: 1968–1988. SOURCE: National Research Council, Office of Scientific and Engineering Personnel data.

As the faculties and graduate student bodies of U.S. universities have become increasingly multinational, new relationships have developed between U.S. research universities and foreign corporations and governments. Ten years ago three-fourths of all research at U.S. universities was financed by federal, state, and local government, with U.S. industry and private foundations providing the balance. Federal funds also contributed very significantly to student financial aid and university faculty improvement. Virtually no American university research or faculty development program was funded by foreign sources, either through research contracts or good will contributions.

In the past 10 years, however, the sources of support for U.S. university research and faculty development have been changing rapidly. The spiraling cost of basic research and rapid expansion of academic research programs have significantly outpaced the growth of federal funding. This, in turn, has forced universities and university-based researchers to cultivate alternative sources of funding. Between 1978 and 1988, the federal government's share of total university research funding shrank from 66 to 60 percent, while the share accounted for by state and local government remained virtually unchanged (down slightly from 8.9 to 8.6 percent). Over the same period, U.S. industry nearly doubled its share from 3.7 to 6.5 percent, while the share of internally generated funding by universities increased from 12 to 18 percent.

In the context of this larger shift in sources of university support, the small but growing contribution of foreign companies and governments to U.S. research universities has attracted considerable attention from U.S. policymakers and the press over the past few years. Although it is widely acknowledged that data on foreign funding of U.S. university research are spotty at best, recent estimates by the U.S. General Accounting Office indicate that, in purely financial terms, foreign support of U.S. university research is trivial—a mere 1 percent of U.S. universities' aggregate research budget, and little more than 4 percent of the research budgets of the top five recipients of foreign funding in 1986.[21] Nonetheless, the results of an informal survey of eight leading U.S. engineering schools, prepared for the committee, indicate that foreign support for university research is already much more of an international enterprise than aggregate financial data alone would suggest.[22]

A more useful measure of the scope and significance of the internationalization of U.S. university research is provided if the definition of "foreign support" is expanded to include nonfinancial as well as financial contributions of foreign entities to U.S. research universities. Viewed from this perspective, foreign support encompasses (1) the participation in university research activities of foreign students, postgraduates, visiting scholars, and research personnel from foreign firms, (2) sponsored and open-ended underwriting of research of foreign corporations and their subsidiaries, (3) the cooperative activities of foreign laboratories set up near U.S. research universities, (4) capital grants of buildings, equipment, and other in-kind contributions by foreigners, and (5) the engagement of U.S. faculty as consultants or advisers by foreign corporations and government agencies.

Of the many different types of foreign support, the contribution of human capital, for the most part independent of foreign corporations and governments, is clearly the most important. Foreign corporations support U.S. university-based research financially and otherwise through a variety of mechanisms: underwriting and supplementing university research personnel by providing scholarships, stipends, and expenses for students and visiting company researchers to work at university laboratories; participating in university industrial liaison programs and university-based interdisciplinary research programs (e.g., Engineering Research Centers and Manufacturing Research Centers); and funding contracts, individually or jointly with other companies or public agencies, for donor-specified research.

In addition to the influx of foreign students and faculty and direct interaction of foreign firms or governments with U.S. universities, there are many other avenues through which U.S. academic research and technical education are becoming increasingly global in orientation and activity. Individual faculty members from U.S. university science and engineering departments frequently consult for foreign firms and governments, and are active participants in international conferences. A number of prominent U.S. research universities are involved in collaborative research efforts with their foreign counterparts. Finally, the growing interest of American undergraduates in study abroad is stimulating another kind of globalization of American colleges and universities, as they establish foreign operations through branch campuses, sister university affiliations, and exchange programs for students and faculty.

NOTES

1. U.S. leadership in total factor productivity has been attributed largely to its leadership in mass production and advanced product technologies. See Nelson (1990).
2. Which data sets provide the most appropriate basis for assessing relative changes in technical capabilities of nations, those which compare absolute values or those comparing ratios such as R&D/GNP, R&D personnel/10,000 workers, output per manufacturing employee? Surely, it is absurd to expect countries with less than half the U.S. population and significantly smaller national material and natural resource endowments than the United States to achieve absolute levels of investment in technical resources (human or financial) on a par with those of the United States. On the other hand, national comparisons of ratios, such as productivity data, and their changes over time offer considerable insight concerning the relative efficiency and effectiveness with which a country employs its basic human, financial, and natural resource endowments, and leverages these endowments through investment in technological innovation.
3. These numbers must be considered only as approximations of R&D employment. First, the categorization of scientists and engineers as R&D personnel varies from country to country; in Japan, only those working full time in R&D are classified as such, whereas the United States calculates "full-time equivalents" of R&D employees. Second, Slaughter and Utterback (1990) calculate the shares of "defense" and "nondefense" R&D personnel by applying the ratios of "defense" to "nondefense" R&D spending to total R&D personnel—admittedly a rough estimate.
4. Defense contracts currently account for nearly one-third of all U.S. industrial R&D. See U.S. Library of Congress (1990, p. 102).
5. The U.S. manufacturing sector employs a quarter of the nation's scientists and engineers, yet it accounts for over 95 percent of what is currently recorded as industrial R&D spending. See National Science Board (1989, pp. 235, 236, 252).
6. Clearly, the yardstick with which one measures the relative technical prowess of a country in one industry need not be the same as that used in another industry, whose products, processes, markets, and technologies differ considerably from the first. For example, in the pharmaceuticals industry, where patenting is pervasive and seen as an effective competitive weapon, the relative distribution of frequently cited patents among national industries may be a useful gauge of overall technical strength. However, in another industry such as petrochemicals where know-how and trade secrets are valued much more as sources of competitive advantage than patents, patenting may prove a poor measure of relative strength.
7. High-tech products are defined by the OECD and U.S. Department of Commerce as products having higher ratios of R&D expenditures to shipments than other product groups. The OECD defines six industries as high-tech following International Standard Industrial Classification (ISIC) codes—drugs and medicines (ISIC 3522); office

machinery, computers (ISIC 3825); electrical machinery (ISIC 383 less 3832); electronic components (ISIC 3832); aerospace (ISIC 3845); and scientific instruments (ISIC 385). The U.S. Department of Commerce, using a more sophisticated R&D tracking technique (DOC-3), defines 10 industries as high-tech following Standard Industrial Classification (SIC) codes: guided missiles and spacecraft (SIC 376); communications equipment and electronic components (SIC 365-367); aircraft and parts (SIC 372); office, computing, and accounting machines (SIC 357); ordnance and accessories (SIC 348); drugs and medicines (SIC 283); industrial inorganic chemicals (SIC 281); professional and scientific instruments (SIC 38 less 3825); engines, turbines, and parts (SIC 351); and plastic materials and synthetic resins, rubber, and fibers (SIC 282). As of 1986, data covered by the OECD high-tech definition equaled that covered by the DOC-3 definition. See National Science Board (1989, pp. 149–150).

8. Note the U.S. high-tech exports as a share of U.S. high-tech production has not changed significantly over the past 20 years, up only 1 percentage point from 10 percent in 1970 to 11 percent in 1986. The sheer size of the U.S. domestic market for high technology and non-high technology products has contributed to a relative neglect of overseas markets by U.S. producers in the past. The urgency of capturing a larger share of non-U.S. markets became apparent only after the relatively insignificant U.S. trade deficits of the 1960s and early 1970s mushroomed with the onset of the oil crises and subsequent import penetration of the U.S. market by the more export dependent producers of Asia and Western Europe. See National Science Board (1989, p. 152).
9. For the classic elaboration of the "product cycle" model, see Vernon (1966).
10. By the late-1960s, U.S. companies, which accounted for 50–60 percent of world outward direct foreign investment in manufacturing at the time, were investing 8–10 percent of their total R&D budgets overseas. Moreover, in a few industries, such as pharmaceuticals and machinery, the flow of technology generated by U.S. overseas subsidiaries back to their U.S. parents was significant. However, in many more industries reverse technology transfer was relatively insignificant, that is, U.S. parent R&D funds were used by most subsidiaries to develop technology for the host market or host region exclusively. The population of European and Asian multinationals remained relatively small into the early 1970s, hence one would expect the transnational R&D activities of European and Asian industry to be even more limited than their U.S. counterpart at the time. For a useful survey of recent trends in reverse technology transfer by U.S. multinationals, see Mansfield and Romeo (1984).
11. Trade, investment, and employment data alone offer only limited insight into the extent of current U.S. economic and technological interdependence. After all, as these data suggest, the vast majority of economic activities in the United States do not involve direct trade of goods or services internationally or direct investment abroad. Only a small fraction of the U.S. services sector (excluding banking) is engaged directly in international trade and investment, although this sector accounted for over 70 percent of U.S. GNP and 75 percent of total U.S. employment in 1986. Similarly, a significant share of U.S. manufacturing is done not by large transnational corporations, but by small- and medium-sized establishments that sell the majority of their output to other U.S.-based firms.

    On the other hand, the extensive interdependence of "domestic" service providers and internationally engaged U.S. manufacturers and service providers is essentially ignored by standard trade and investment data. Moreover, the share of U.S. manufacturing involved in supplying components, materials, capital goods, and other intermediate products to transnational companies or their first and second tier suppliers is surely considerably greater than trade figures alone would suggest.
12. It is estimated, for example, that IBM and Hewlett-Packard do nearly 30 percent of their R&D work outside of the United States. It is also interesting to note that U.S.

firms' sales of technology to foreigners through licensing agreements increased noticeably during the 1980s. According to the U.S. Department of Commerce, U.S. receipts from such technology sales increased from $1.4 billion in 1980 to $2.1 billion in 1987 (1982 prices). Japan was the largest consumer of U.S. technology sold through these agreements, accounting for 41 percent of all U.S. royalty and licensing fee receipts in 1987 (National Science Board, 1989).

13. U.S. Department of Commerce, (1984, 1985a, 1985b, 1986, 1987, 1988, 1988a), Table H-3 Research and Development Expenditures by Affiliates, by Industry of Affiliate; National Science Foundation (1989, Table B-11, p. 27). Admittedly, data regarding R&D expenditures say very little about the nature of advanced technical activities of a firm. Foreign firms are accused of setting up research tracking or technology transfer operations in the United States and labelling them as research and development. Conversely, U.S. firms operating overseas may label activities only vaguely related to R&D as such in an effort to comply with domestic content laws.
14. According to Hagedoorn and Schakenraad (1990a), biotechnology includes "relevant basic research and all applications of that particular field of technology in agriculture, pharmaceuticals, ecology, nutrition, chemicals and basic research. Information technologies are confined to computers, industrial automation, microelectronics, software and telecommunications. New materials are defined as new and improved electronics materials, technical ceramics, fibre-strengthened composites, technical plastics, powder metallurgy and special metals and alloys."

    The authors identify six major modes of technology cooperation for the technologies surveyed: joint R&D, joint ventures, technology exchange agreements, cross-equity holdings, customer-supplier relations, and one-directional technology flows. While joint R&D represents the leading mode of collaboration in all three technology fields (25–30 percent of total), the relative importance of other modes of cooperation varies significantly among the three fields. Direct investment figures prominently in biotechnology, whereas one-directional flows and joint ventures are more prevalent in information technology and new materials. See Hagedoorn and Schakenraad (1990a, p. 3).
15. Applying several analytical techniques, the authors demonstrate the structure of intercorporate technical networks, the clustering of interfirm alliances within the networks, and the changing density of these networks over time. Their model only accommodates a limited number of companies, i.e., the 45 companies involved in the most alliances. See Hagedoorn and Schakenraad (1990a, pp. 22a-b).
16. Following the classification scheme for "high-technology industries" developed by Riche, Hecker, and Burgan (1983, pp. 52–53), we define "engineering-intensive industries" to include (a) those in which the ratio of R&D to net sales is equal to or greater than two times the average for all industries; or (b) those in which the ratio of technology-oriented workers (engineers, life and physical scientists, mathematical specialists, engineering and science technicians, and computer specialists) to total work force is at least one and a half times the average for all industries; or (c) those in which the ratio of technology-oriented workers to total work force is equal to or greater than the average for all manufacturing industries *and* the R&D/sales ratio is close to or above the average for all industries.
17. "Biotechnology is not an industry *per se*, but rather an array of technologies that can be applied to a number of industries. These technologies include: molecular and cellular manipulation, enzymology, X-ray crystallography, computer modeling, biomolecular instrumentation, industrial microbiology, fermentation, cell culturing, and separation and purification technologies." U.S. Department of Commerce (1989a, p. 19-1).
18. The committee's working definition of technology includes both the generation of new products or services and the associated organizational and managerial know-how, such as "just-in-time" production systems, quality circles, and "total quality control." To be

sure, the criteria for selecting the most critical technologies or technology areas for an entire industry or industry subset are multiple, complex, and ultimately highly subjective, i.e., based on the best judgment of the committee, which has been informed, in turn, by the comments and advice of numerous industry experts in the United States and abroad (see Appendix A for individual industry profiles).

19. The causes of this decline in U.S.-born enrollments in engineering doctoral programs remain the subject of debate. Nonetheless, it is worth noting that the falloff in U.S. nationals' graduate enrollments appears to track the decline in federal fellowship (not research assistantship) support for graduate study in the early 1970s.
20. So far, U.S. engineering schools have had few problems recruiting foreign talent, most of it from newly industrializing or developing countries such as Taiwan, Korea, the People's Republic of China, and India. Foreign students and faculty from these countries are clearly attracted to U.S. universities by the quality of their research facilities, the reputation of their faculty and graduates, and their access to the lucrative U.S. job market. Also, one should not underestimate the drawing power of U.S. political, religious, and social freedoms for students from countries with less tolerant political and social regimes. U.S. graduate schools are further assisted in their search for foreign talent by foreign governments, which, in an effort to build their own technological infrastructures, encourage their nationals to study or pursue postdoctoral research at U.S. universities before returning home to work.
21. Although some public institutions are required by state law to report foreign funds, most of them have not successfully differentiated between domestic and foreign financial support. Admittedly, it is not at all obvious how one would categorize contributions of foreign alumni, or those of a U.S. subsidiary of a foreign company or a U.S. company's foreign subsidiary. However, even without attempting the foreign versus domestic distinction, the lack of uniform university accounting procedures, the multiplicity of funding sources and channels, and the decentralized nature of exchanges between donors and a broad spectrum of university offices, departments, and individual researchers, all contribute to make the tracking of foreign support extremely haphazard. See National Science Foundation (1989b) and General Accounting Office (1988).
22. In December 1989, the National Academy of Engineering, as part of this study, helped sponsor the research of Helena Stalson on foreign support of U.S. university-based research. Stalson's draft report, "Foreign Participation in Engineering Research at U.S. Universities," is based on interviews conducted at eight universities during the spring of 1989: Carnegie Mellon, Columbia, Cornell, Massachusetts Institute of Technology, Princeton, Rensselaer Polytechnic Institute, University of Illinois (Urbana), and University of Wisconsin (Madison).

REFERENCES

Council on Competitiveness. 1990. Competitiveness Index 1990. Washington, D.C.: Council on Competitiveness.

Chenais, Francois. 1988. Multinational enterprises and the international diffusion of technology. Pp. 496–527 in Technical Change and Economic Theory, G. Dosi, C. Freeman, R. Nelson, G. Silverberg, and L. Soete, eds. London: Pinter Publishers Ltd.

Economic Report of the President. 1990. Transmitted to the Congress, February 1990. Washington, D.C.: U.S. Government Printing Office.

Enderwick, Peter. 1989. Multinational corporate restructuring and international competitiveness. California Management Review (Fall):44–58.

Graham, Edward M., and Paul R. Krugman. 1989. Foreign Investment in the United States. Washington D.C.: Institute for International Economics.

Hagedoorn, John, and Jos Schakenraad. 1989. Strategic Partnering and Technological Co-operation. Maastricht Economic Research Institute on Innovation and Technology. The Netherlands: MERIT.
Hagedoorn, John, and Jos Schakenraad. 1990a. Leading Companies and the Structure of Strategic Alliances in Core Technologies. Maastricht Economic Research Institute on Innovation and Technology. The Netherlands: MERIT.
Hagedoorn, John, and Jos Schakenraad. 1990b. Inter-firm Partnerships and Corporate Strategies in Core Technologies. Maastricht Economic Research Institute on Innovation and Technology. The Netherlands: MERIT.
International Monetary Fund. 1990. World Economic Outlook. April.
Julius, DeAnne. 1990. Global Companies and Public Policy: The Growing Challenge of Foreign Direct Investment. New York: Council on Foreign Relations.
Mansfield, Edwin, and Anthony Romeo. 1984. "Reverse" transfers of technology from overseas subsidiaries to American firms. IEEE Transactions on Engineering Management EM-31 3:122–127.
Mowery, David C., ed. 1988a. International Collaborative Ventures in U.S. Manufacturing. Cambridge, Mass.: Ballinger Publishing Company.
Mowery, David C. 1988b. Joint ventures in the U.S. aircraft industry. Pp. 71–110 in International Collaborative Ventures in U.S. Manufacturing, D.C. Mowery, ed. Cambridge, Mass.: Ballinger Publishing Company.
National Research Council. 1988. Foreign and Foreign-Born Engineers in the United States. Office of Scientific and Engineering Personnel. Washington, D.C.: National Academy Press.
National Science Board. 1989. Science and Engineering Indicators - 1989. Washington, D.C.: U.S. Government Printing Office.
National Science Foundation. 1988. International Science and Technology Update: 1988. Research and Development (R&D) Expenditures. Special Report NSF 89-307, Detailed Statistical Tables. Washington, D.C.: National Science Foundation.
National Science Foundation. 1989a. Research and Development in Industry: 1987. Surveys of Science Resources Series NSF 89-323, Detailed Statistical Tables. Washington, D.C.: National Science Foundation.
National Science Foundation. 1989b. Report of the NSB Committee on Foreign Involvement in U.S. Universities. Washington, D.C.: National Science Foundation.
Nelson, Richard R. 1990. U.S. technological leadership: Where did it come from and where did it go? Research Policy 19:117–132.
Riche, Richard W., Daniel E. Hecker, and John U. Burgan. 1983. High technology today and tomorrow: A small slice of the employment pie. Monthly Labor Review (Nov):50–58.
Slaughter, Sarah, and James Utterback. 1990. U.S. research and development: An international comparative analysis. Business in the Contemporary World (Winter):27–35.
United Nations Centre on Transnational Corporations. 1988. Transnational Corporations in World Development: Trends and Prospects. New York: United Nations.
U.S. Department of Commerce. 1984. Foreign Direct Investment in the United States: Annual Survey Results, Revised 1981 Estimates. Bureau of Economic Analysis. Washington, D.C.
U.S. Department of Commerce. 1985a. Foreign Direct Investment in the United States: Operations of U.S. Affiliates of Foreign Companies, Revised 1982 Estimates. Bureau of Economic Analysis. Washington, D.C.
U.S. Department of Commerce. 1985b. Foreign Direct Investment in the United States: Operations of U.S. Affiliates, 1977-80. Bureau of Economic Analysis. Washington, D.C.
U.S. Department of Commerce. 1986. Foreign Direct Investment in the United States: Operations of U.S. Affiliates of Foreign Companies, Revised 1983 Estimates. Bureau of Economic Analysis. Washington, D.C.

U.S. Department of Commerce. 1987. Foreign Direct Investment in the United States: Operations of U.S. Affiliates of Foreign Companies, Revised 1984 Estimates. Bureau of Economic Analysis. Washington, D.C.

U.S. Department of Commerce. 1988a. Foreign Direct Investment in the United States: Operations of U.S. Affiliates of Foreign Companies, Revised 1985 Estimates. Bureau of Economic Analysis. Washington, D.C.

U.S. Department of Commerce. 1988b. Foreign Direct Investment in the United States: Operations of U.S. Affiliates of Foreign Companies. Bureau of Economic Analysis. Preliminary 1986 Estimates. Washington, D.C.

U.S. Department of Commerce. 1988c. International Direct Investment: Global Trends and the U.S. Role. International Trade Administration. Washington, D.C. November.

U.S. Department of Commerce. 1989a. U.S. Industrial Outlook 1989. Washington, D.C.

U.S. Department of Commerce. 1989b. Staff Report—Direct Investment Update: Trends in International Direct Investment. International Trade Administration, Office of Trade and Investment Analysis. Washington, D.C. September.

U.S. Department of Labor. 1990. Comparative Real Gross Domestic Product, Real GDP Per Capita, and Real GDP Per Employed Person, Fourteen Countries, 1950–1989. Bureau of Labor Statistics, Office of Productivity and Technology. April. Unpublished data.

U.S. General Accounting Office. 1988. R&D Funding: Foreign Sponsorship of U.S. University Research. Report prepared for the Honorable Lloyd Bentsen, U.S. Senate. Gaithersburg, Md.: U.S. General Accounting Office.

U.S. Library of Congress, Congressional Research Service. 1990. Japan-U.S. Economic Issues: Investment, Saving, Technology and Attitudes. Japan Task Force Report No. 90-78 E. Washington, D.C.

Vernon, Raymond. 1966. International investment and international trade in the product cycle. Quarterly Journal of Economics 80:90–207.

Vonortas, Nicholas S. 1989. The Changing Economic Context: Strategic Alliances Among Multinationals. Center for Science and Technology Policy. Rensselaer Polytechnic Institute, Troy, N.Y. February.

Womack, James P. 1988. Multinational joint ventures in motor vehicles. Pp. 301–348 in International Collaborative Ventures in U.S. Manufacturing, D.C. Mowery, ed. Cambridge, Mass.: Ballinger Publishing Company.

2

Opportunities and Challenges of Globalization

THE PROMISE OF GLOBALIZATION

The globalization of technical activities and the closing of the postwar technology gaps among nations offer both the United States and its trading partners several major opportunities for technical and economic advance. First, **the globalization of industry and technology promises to accelerate transnational integration and cross-fertilization in engineering, technology, and management**. As multilateral flows of trade, investment, and technology increase and more companies are drawn into global industrial networks of production, research, finance, and distribution, more firms are able to exploit the special competencies and technologies of an ever larger number of world-class national technical enterprises. This, in turn, speeds the development and diffusion of new product and process technologies and new "best practice" engineering and management techniques worldwide.

Thirty years ago, international technology flows between U.S. corporations and their foreign affiliates in most high-technology industries were by and large unidirectional, outward from the United States. Since the mid-1960s, the flow of technology and know-how between American and affiliated or unaffiliated foreign firms has become increasingly reciprocal as foreign technical competence has grown (Mansfield and Romeo, 1984). During the 1980s, however, both the pace of reverse technology flows into the United States and public appreciation of its significance increased rapidly.

This is perhaps most apparent in technologically more mature U.S. industries, such as steel and automobiles, where foreign, particularly Japanese, product and process technologies and management techniques have made significant contributions toward improving overall performance in recent years. Yet even the most technically dynamic industries studied by the committee demonstrated a shift toward more reciprocal flows of technology and "best practice" engineering and management techniques. Consider, for example, relatively recent adoption by U.S. companies such as Motorola, Xerox, and Hewlett-Packard of techniques developed by Japanese firms for the management of technology and other productive resources, for example, total quality control, just-in-time manufacturing, and concurrent engineering. The contribution of Japanese and European companies to the advance of specific product and process technologies in high-technology industries is clearly demonstrated by both patent and trade data (see Figure 1.6 above), and the committee's case studies of the aircraft engine, computer printer, and semiconductor industries (see Appendix A).

Second, **competitive globalization of technical activities promises to enhance the diversity and depth of the current stock of world engineering and scientific resources and thereby provide greater stimulus to economic growth and technology development**. In the context of competitive, open markets, global sourcing, assembly, production, and research permit private corporations to increase the efficiency with which they employ technical resources. In an increasingly integrated global economy, firms are able to access larger markets, a larger pool of specialized technical competence, and a larger reserve of complementary assets such as managerial talent, capital, and skilled labor. This, in turn, offers them the opportunity to increase economies of scale and scope across the spectrum of technical functions encompassing research, design, development, production, sales, and service.

In addition to increasing the efficiency with which technical resources are applied in advanced industrialized economies, the global development and acquisition of technology and know-how by corporations promise to integrate a growing number of less-developed national technical enterprises into emerging global industrial technology networks. As large numbers of highly trained, low-cost engineers and scientists in countries such as India, the People's Republic of China, Indonesia, the German Democratic Republic, Czechoslovakia, and Hungary become increasingly linked with global product, service, and factor markets, the productive potential of available human assets is certain to increase.[1]

Without hard data concerning world demand and the price elasticity thereof for science and engineering services, it is impossible to state conclusively whether the anticipated increase in productivity of the globe's technical work force would raise or lower total world demand for engineers and scientists. Nevertheless, there are numerous factors that cause the committee to believe that the growth of world demand for all sorts of technical talent will continue to outstrip the growth of even a more productive world supply well into the next century.

The documented secular shift in patterns of consumption toward increasingly technology-intensive products and services in the United States and other advanced industrialized countries should provide a sustained boost to demand for sophisticated technical services, particularly in the areas of information technology, biotechnology, and advanced materials. The advanced age and poor condition of public infrastructure, such as transportation systems, energy systems, water and waste treatment facilities, and housing, in many advanced industrialized countries, not to mention the lack of these vital infrastructures in many industrializing countries should also place major demands on a wide range of engineering and scientific talent in coming decades. Finally, the magnitude and intractable character of current global environmental problems, such as global warming or solid and hazardous waste reduction and disposal, are certain to require vast human technical resources to develop and apply tools, concepts, and specific technologies to meet these challenges.

CHALLENGES FACING THE UNITED STATES AND ITS TRADING PARTNERS

Despite the lack of conclusive quantitative evidence, it is the best judgment of the committee that increased global integration of national technical enterprises will contribute to world economic growth, technical advance, and world demand for science and engineering services in coming decades. At the same time, it must be acknowledged that the process of globalization will continue to involve a spatial redistribution of industrial and associated technical activities and will necessarily benefit some countries and companies more than others. In short, although the rising tide of engineering competence worldwide and the greatly increased transnational mobility of technology and other factors of production promise greater efficiencies and economic growth in the aggregate, they have also intensified and recast competition among firms and nations in the process.

Inward-looking corporate and national strategies for economic competitiveness, strategies preoccupied with the management of essentially indigenous markets, technology, and other factors of production, are being rendered ineffective or irrelevant by the process of globalization. As a result, nations as well as firms are being forced to recognize that it is no longer possible to achieve or sustain leadership in technologies vital to their future competitiveness and economic growth without greatly improving both the firm's and the nation's ability to capture a fair share of the benefits of what is becoming a truly global technical enterprise. Unfortunately, major obstacles to the full exploitation of the global technical enterprise are emerging as the process of globalization gathers momentum.

Forces That Lead to Domestic Protectionist Response

The globalization of industry and technology creates new winners and losers within a national economy. Some industries or regions experience growth as a result of integration into global markets, while others, unable to weather the force of global competition, undergo economic decline. For example, the U.S. aircraft, computer, and telecommunications industries and their host locations have watched their business opportunities expand dramatically with the globalization of markets for their products. In contrast, the U.S. machine tool, steel, and automotive industries and their host communities have experienced severe economic dislocation and contraction as a result of growing foreign competition over the past 10 to 15 years. The problem is that those who benefit from globalization tend to take their good fortune for granted or attribute it entirely to their own superior efforts and initiatives. The losers tend to blame their losses on an unfair and hostile world, from which they demand protection. Hence, the winners often fail to appreciate their stake in globalization, while the losers are fully conscious that globalization is the source of their problems. This difference in the perceived stakes of globalization creates a domestic political imbalance that often fosters protectionism. Another source of protectionism is the fact that the costs of protective actions are widely distributed and the benefits are highly concentrated among regions or industries. This gives the beneficiaries of protection considerably larger incentives for political mobilization than it does to the general public, that is, consumers and taxpayers, who usually pay the price.

International Asymmetries of Market Access

Some national economies are more closed to the reciprocal flow of technology, trade, and investment across their borders than others. One need only consider the patterns of trade and foreign direct investment among the world's technologically most dynamic economies to appreciate this fact. International comparison of average levels of intraindustry trade—the extent to which a nation exports and imports similar products—offers one particularly illustrative perspective on the anomaly that is Japan.

A country's level of intraindustry trade is suggestive not only of the relative specialization and sophistication of its industrial base but also of the degree to which its economy is open to exports from its industrial and technological peers. In this context, the greater a nation's ratio of intraindustry to total trade, the more "open" its economy is to the products and services of its industrialized trading partners. As Table 2.1 suggests, however, over the postwar period as intraindustry trade has greatly expanded as a share of total trade, Japan has consistently registered average levels of intraindustry trade that are out of line with those experienced by other industrialized nations (Lincoln, 1990).

In the area of foreign direct investment, the anomalous status of Japan is equally apparent (Figure 2.1). Western Europe has absorbed more than a third of total world direct investment for decades. The United States, despite accounting for a relatively small share of world inward investment for most of the period since 1945, has watched its share increase to more than one-fourth of the world total during the past decade. Japan, on the other hand, continues to account for a remarkably small share of total inward investment, even though its share of total outward direct investment has grown rapidly in recent years.

The reasons behind the more mercantilist character of particular nations are complex, having as much to do with the timing, historical context, and structural consequences of a country's industrialization, its culture, or its legal traditions, as with its specific public policies. Regardless of its causes, however, differential national treatment of international trade, technology, and investment flows has contributed to bilateral economic imbalances, increased international political friction, and fostered protectionism. In so doing, it has made the task of adjustment to the economic challenges and opportunities of globalization more difficult for all nations, although particularly for more open national economies.

The Different "Learning" Aptitudes of Nations

Some nations are much better at taking advantage of a globalizing technology base than others. Free access to another country's markets, technologies, and financial resources is of limited use to a nation if its citizens and corporations are unwilling or unable to take advantage of the opportunity. During the nineteenth and early twentieth centuries, the United States made its successful bid for global industrial leadership by rapidly assimilating and improving upon technologies and techniques first developed in other countries. Following World War II, however, the need, and with it, the ability of U.S.-based companies to assimilate and exploit foreign technology and know-how seems to have declined markedly relative to that of its main trading rivals, most notably Japan.

At a time when the sources of technical advance in a growing number of industries are becoming more widely dispersed throughout the globe, the "not-invented-here" syndrome, a product of decades of unchallenged U.S. technological supremacy, poses a severe handicap to the country.[2] Meanwhile, Japan has assumed the former American role as the industrial-

TABLE 2.1 Average Intraindustry Trade, Five Countries, Selected Years: 1959–1985. Intraindustry trade index points.[a]

	Index Basis						
	Three-digit SITC[b] categories						Four-digit SITC categories
Country	1959	1964	1970	1975	1980	1985	1985
All traded products							
Japan	17	21	26	19	19	23	—
United States	40	40	53	57	57	54	—
France	45	60	67	65	67	74	—
West Germany	39	42	54	52	57	63	—
Manufactured products only							
Japan	—	—	32	26	28	26	23
United States	—	—	57	62	62	61	54
France	—	—	78	78	82	82	74
West Germany	—	—	60	58	66	67	63
South Korea	—	—	19	36	40	49	44

SOURCE: Lincoln (1990, p. 47). Reprinted with permission.

[a] The calculation of intraindustry trade (IIT) in a single industry is based on the standard equation, $IIT_i = [1-[x_i-m_i]/[x_i+m_i]] \times 100$, where i = industry, x = exports, and m = imports. The average index for trade in all industries within a nation is calculated by weighting each industry by its share in total trade.

[b] SITC = Standard International Trade Classification

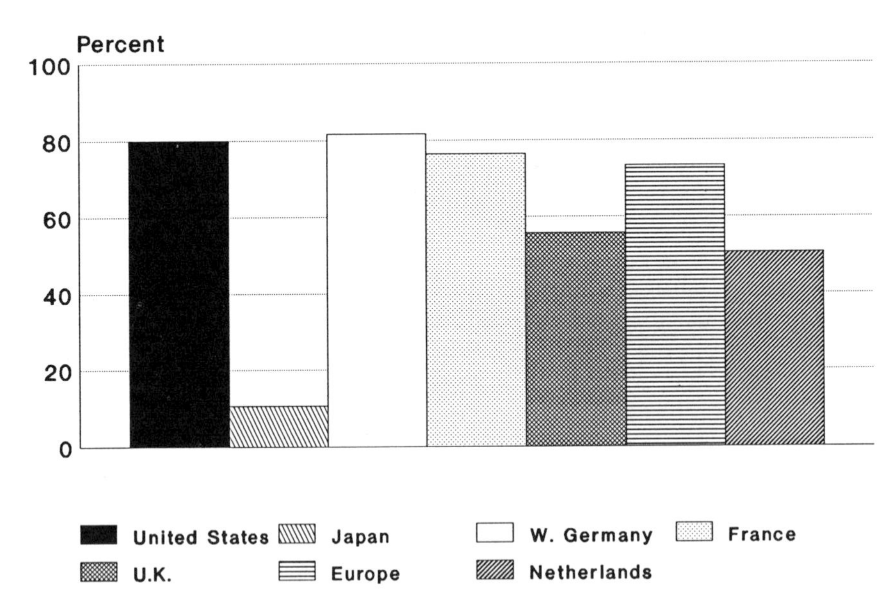

FIGURE 2.1 Ratio of inward to outward stocks of foreign direct investment, by selected countries: 1987. SOURCE: U.S. Department of Commerce (1989, p. 11, p. 15).

ized world's most diligent student of other nations' technical practices, inviting both the admiration of those impressed by its diligence and the wrath of those who view the continuance of such behavior as the most insidious form of "free-riding" possible.

Again, it is important to understand that the different "learning" aptitudes of nations, like differences in the relative openness of their economies, stem as much from the structural, institutional, legal, and cultural consequences of their unique political and economic development as from particular public policies. Hence there is no quick or simple policy response that will eliminate the learning differential. The result, however, is an international technological order in which there are additional impediments to reciprocal transfers of technology and know-how. Though not explicitly "protectionist," these impediments generate additional political tension between nations and often encourage policy responses that impede political and economic adjustment to new global realities.

The Threat of Global Monopolies

The recent surge in national and transnational mergers, takeovers, and strategic alliances in highly concentrated industries such as the production of electrical equipment, computers, semiconductors, automobiles, and aircraft engines underlines the inherent contradictions of corporate strategies and public[3] policies with regard to "competitiveness" and "competition." Much of the recent transnational alliance activity among erstwhile competitors has involved major companies in industries such as semiconductors and aircraft, where the sheer technological complexity, high initial capital costs, and spiraling cost of technological advance already pose virtually insurmountable barriers to market entry. At the present time, these companies' alliances, joint ventures, and cross-licensing agreements do not appear to be anticompetitive in motive or consequences. However, it is not unreasonable to anticipate that some of these alliances will eventually impede competition, with negative consequences for economic growth and technical advance (Porter, 1990).

In other technologically more mature concentrated industries, anticompetitive behavior becomes more of a possibility as markets for certain products mature or saturate. In a mature market, incremental improvements in product technology tend to become more and more costly, so that firms are increasingly tempted to forgo such improvements in order to preserve their profitability by avoiding an increasingly cost-ineffective product improvement race.

It is important to recognize that the potential for harmful anticompetitive behavior by global companies has been exacerbated by the trade, technology, and industrial policies of the industrialized nations in recent years.

"Managed trade" agreements such as the 1986 U.S.-Japan Semiconductor Agreement often encourage cartel-like behavior. Anticompetitive behavior may also be reinforced by the "closed" technology development programs funded by the governments of many industrialized countries.

At the same time, there is growing evidence that national competition policies or antitrust laws are becoming significant obstacles to cross-border mergers and acquisitions that do *not* undermine national, regional, or global competition (Julius, 1990). Such policy-induced impediments to international competition in the name of antitrust enforcement also threaten to undercut economic growth and technical advance.

GLOBALIZATION: ON BALANCE A POSITIVE TREND

On balance, the committee is convinced that the globalization of R&D, production, investment, markets, and technology is a positive trend for both the United States and the rest of the world. Most important, globalization of technical activities represents a trend that cannot be reversed or significantly impeded by national governments without inflicting high costs on their citizens. At the same time, the committee recognizes that failure to advance effective domestic policies and international negotiations toward the objective of reducing impediments to the competitive globalization of industry and technology is likely to encourage protectionist policies by governments of the advanced industrialized nations. Whether policy obstacles to the free movement of goods, capital, labor, and technology accumulate gradually or explode in trade or investment wars, they are bound to increase global economic dislocation, delay needed structural adjustment, and impede economic growth and technical advance for all nations. Either scenario, although costly to the advanced industrialized nations, would have particularly harsh consequences for developing and newly industrializing countries.

NOTES

1. For example, consider the rapid pace at which the once isolated technical work force of Eastern Europe is being drawn into the global technical order through the recent actions of European, Asian, and North American multinationals in the automotive, electrical equipment, and chemical industries.
2. Given the recent changes in the distribution of world technology-intensive trade, the rapid growth of non-U.S. foreign direct investment, and the declining share of total patents granted U.S. citizens, perhaps the persistently large U.S. technological balance of payments surplus (royalty receipts minus royalty payments) should be interpreted not so much as a sign of technological "free-riding" by our major competitors as an indicator of the U.S. relative inability to absorb foreign technologies.

REFERENCES

Julius, DeAnne. 1990. Global Companies and Public Policy: The Growing Challenge of Foreign Direct Investment. New York: Council on Foreign Relations Press.

Lincoln, Edward J. 1990. Japan's Unequal Trade. Washington, DC: The Brookings Institution.

Mansfield, Edwin, and Anthony Romeo. 1984. "Reverse" transfers of technology from overseas subsidiaries to American firms. IEEE Transactions on Engineering Management EM-31(3):122-127.

Porter, Michael E. 1990. The Competitive Advantage of Nations. New York: The Free Press.

U.S. Department of Commerce. 1989. Staff Report—Direct Investment Update: Trends in International Direct Investment. International Trade Administration, Office of Trade and Investment Analysis. Washington, D.C. September.

3

Strengths and Weaknesses of the U.S. Technical Enterprise

In addition to presenting many challenges and opportunities to the United States, the globalization of technical activities has underlined the special strengths and weaknesses of the U.S. technical enterprise. These must be fully appreciated and their interrelationship better understood before exploring the policy implications of a global technical enterprise.

U.S. COMPARATIVE STRENGTHS

The emergence of world-class technical enterprises abroad and the accompanying increase in global competition demand that the United States take greater stock of its areas of strength in order that they may be more fully developed and exploited. In this regard, a short list of the most important of U.S. national assets should include: a formidable basic research enterprise; a superior advanced technical education system; a vast, technologically demanding domestic market; a large, cosmopolitan, and highly skilled technical elite; and an educational system that fosters individual creativity and inventiveness.

The National Research Enterprise

The U.S. basic research enterprise is unsurpassed. Although the industrialized and industrializing economies of Europe and Asia have expanded their basic research efforts more rapidly than the United States during the last 40 years, the United States retains an impressive, absolute lead in terms of money spent on basic research, the number of scientists and engineers engaged therein, and the volume and quality of basic research output. As Figures 3.1 and 3.2 show, by the late 1980s, the United States was still spending almost as much as Japan, West Germany, France, and the United Kingdom combined on research and development, and had nearly twice as many R&D scientists and engineers as its closest competitor, Japan. Similarly, comparisons of the leading industrialized nations' shares of world patents and scientific literature (Figures 3.3 and 3.4) provide a window on the continuing leadership of the United States in the overall output of pure research.

At the same time, the structure of the U.S. basic research enterprise makes it easily accessible to foreign firms and governments. To begin with, the U.S. university system, which thrives on openness and the free currency of ideas, plays a central role in the U.S. basic research effort. Furthermore, the United States enjoys the world's most extensive and efficient infrastructure for the dissemination of basic research through conferences, technical associations, and technical journals. In short, the very "cosmopolitan" ethos and commitment to the free flow of ideas that contribute so effectively to the extraordinary vitality and productivity of the U.S. basic research enterprise make it extremely difficult for U.S. firms or the United States as a nation to appropriate exclusively the enterprise's basic research product (U.S. General Accounting Office, 1988a,b).

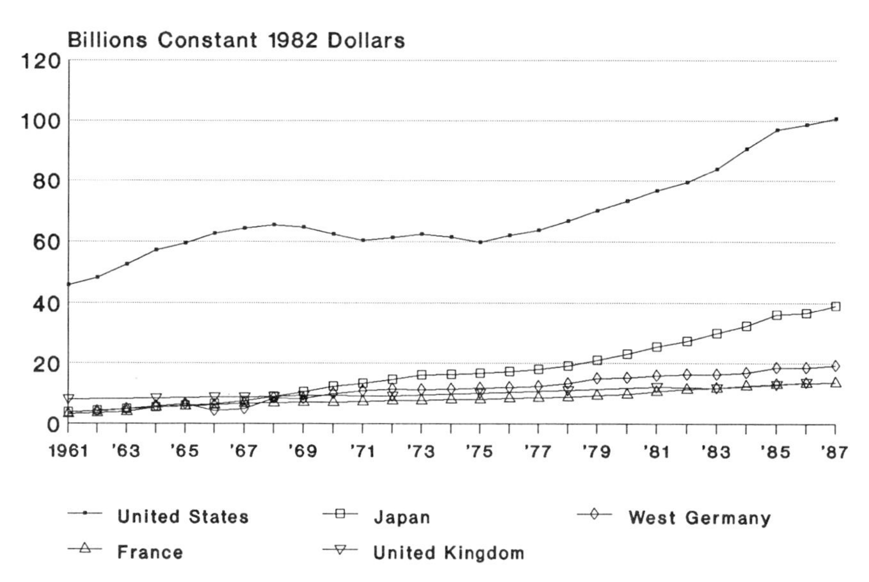

FIGURE 3.1 National R&D expenditures, by selected countries: 1961–1987. SOURCE: National Science Foundation (1988, p. 4).

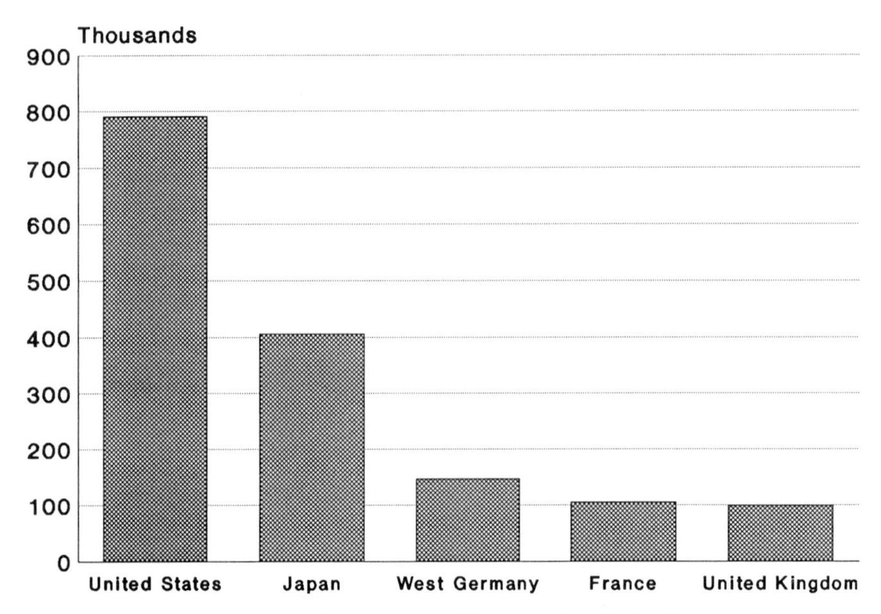

FIGURE 3.2 Scientists and engineers engaged in research and development, by country: 1986. SOURCE: National Science Foundation (1988, p. 36).

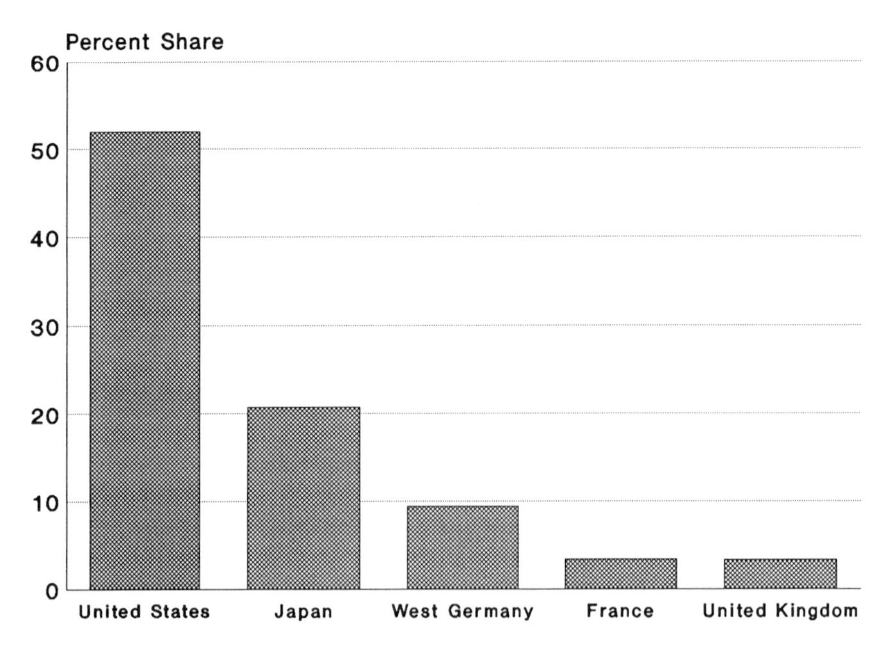

FIGURE 3.3 National shares of patents granted in the United States, by country of residence of inventor and year of grant, all technologies: 1988. SOURCE: National Science Board (1989, p. 362).

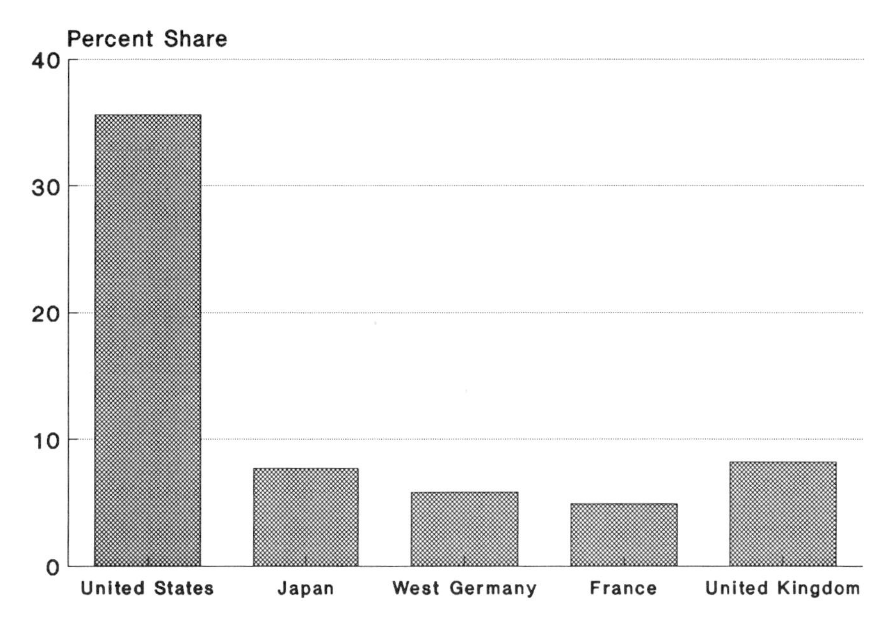

FIGURE 3.4 Shares of world scientific literature, by country: 1986. SOURCE: National Science Board (1989, p. 331).

Advanced Technical Education

The superiority of advanced technical education in America is recognized throughout the world, as attested by persistently large enrollments of foreign students in doctoral level engineering and science programs in American universities. U.S. university engineering and science faculties have long educated many of the best and brightest graduate students from all over the world. Indeed, the attraction of graduate engineering study in the United States is a function of many factors, including the high quality of U.S. university research facilities, the reputation of their faculty and graduates, and the prospect of more rewarding employment in the United States upon graduation, not to mention the drawing power of U.S. political, religious, and social freedoms. The particular strength of U.S. advanced technical education owes a great deal to the fact that U.S. universities have assumed a central role in the nation's basic research enterprise since World War II. As a result, U.S. universities command a large share of the country's total research budget.

The Domestic Market

The relative scale, homogeneity, depth, and openness of the U.S. domestic market have proven a powerful engine for innovation and its commercialization. In addition to being the world's largest single market with per capita purchasing power greater than all but a few of its trading partners, the U.S. domestic market remains the most technically demanding in the world.[1]

Over the past 20 years, the United States has consistently consumed between 40 and 50 percent of world output of high-technology products—more than Japan, France, West Germany, and the United Kingdom combined (Figure 3.5). Likewise, comparatively high U.S. per capita consumption of technically advanced products such as televisions, VCRs, personal computers, facsimile machines, and cellular telephones also attest to the overall technical sophistication of the U.S. market.

Although rapid advances in information and communication technologies have reduced somewhat the technical and organizational advantages of proximity to market in many industries, the sheer size and technology "pull" of domestic demand continue to make the United States an attractive place for firms of all nationalities to design, develop, and market new products, services, and technologies. In this regard, the relative efficiency and size of the U.S. services sector is a major driver of technological advance and innovation in both services and manufacturing industries. Accounting for more than 75 percent of U.S. employment and 71 percent of U.S. GNP, U.S. services industries, which include transportation, communication, health care, and business services, among others, are the world's largest and most efficient. They are also major drivers of technology development and commercialization both as consumers of technology-intensive goods and services and as service providers. Furthermore, the unrivaled scope and dynamism of

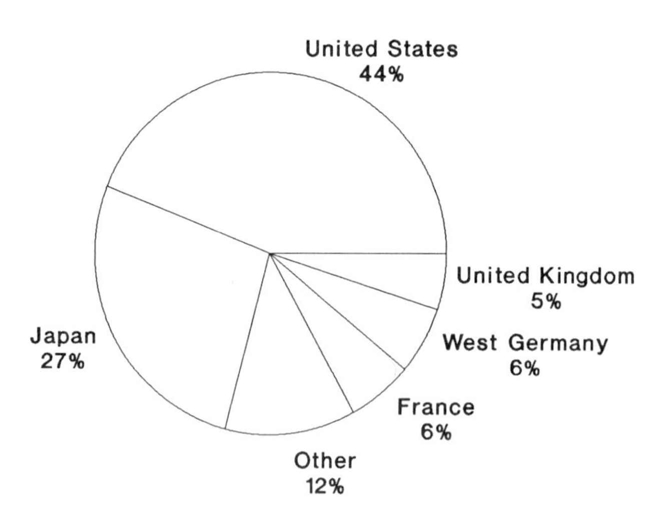

FIGURE 3.5 Home markets for high-technology products, by selected countries: 1986. SOURCE: National Science Board (1989, p. 373).

the U.S. venture capital market has made the United States the world's "mecca" of high-tech entrepreneurs more generally.[2]

Finally, the freedom of the U.S. domestic markets for goods, services, capital, labor, and technology allows productive resources to move more readily from one sector to another in response to changes in demand than they do in most advanced industrialized countries. This allocative efficiency along with the relative openness of the U.S. economy to foreign products, services, and investment, has contributed significantly to the technological dynamism of domestic markets.[3]

It should be noted, however, that the openness of the U.S. economy to foreign imports and investment has extended the benefits of the large, homogeneous, technically dynamic U.S. market to firms from competing nations as well as to "indigenous" companies.

Information Technology

The relative strength of the United States in information technology, especially applications and systems software, expert systems, and artificial intelligence, affords a significant potential advantage in the efficient delivery of sophisticated engineering services. In addition to hosting the world's largest population of computer specialists, the United States has long led the rest of the world in per capita consumption of personal computers, workstations, and software. Currently there are more supercomputers in operation in the United States than in any other country. Furthermore, the United States enjoys a considerable advantage over its competitors in the development and application of expert systems. It is estimated that in 1989 alone, 4,400 expert systems were installed in the United States. In addition, linking these vast hardware and software resources is the world's most extensive, technologically advanced, competitively priced telecommunications system, which includes high-performance networks such as INTERNET (U.S. Bureau of the Census, 1988; U.S. Department of Commerce, 1989a).

This advantage may be partially offset, however, by the relatively easy exportability of software production capabilities to low-wage countries with an underutilized engineering work force. The limited scope and weak enforcement of current international intellectual property laws also make it relatively easy for unscrupulous parties abroad to steal certain types of information technology outright.

The Nation's Pool of Technical Talent

The size and diversity of the nation's pool of technical talent are unmatched by any of the U.S. trading partners. U.S.-based corporations are able to draw on a science and engineering "melting pot" that has been enriched by the infusion of a wide variety of cultures, intellectual traditions, and technical practices. The contributions of successive waves of technically trained immigrants from Europe, Asia, and other parts of the world to U.S. science and engineering are everywhere apparent. Naturalized citizens figure prominently among the ranks of U.S. Nobel Laureates, the National Academies of Sciences and Engineering, and other honorary scientific and technical societies. As of the early 1980s, more than one-sixth of the U.S. engineering work force was foreign-born as was nearly one-half of the U.S. engineering faculty under the age of 36 (National Research Council, 1988).

Indeed, there are economic, social, and political costs associated with high levels of ethnic diversity in a single nation. Racial, linguistic, cultural, and religious differences often compound socioeconomic divisions in the United States, making the politics of education, employment, and resource allocation more contentious than they might be in an ethnically more homogeneous society. Despite these liabilities, however, the committee views the ethnic pluralism of the United States as a major source of strength for the U.S. technological enterprise, as well as for the U.S. political and economic systems.

The Cultivation of Individual Creativity and Initiative

Despite its many failings, the U.S. educational system and the political values that undergird it cultivate individual creativity and individual initiative to an extent far greater than those of other countries. Although the average math and science scores of American high school students remain well below those of their counterparts in Western Europe and Asia, the best U.S. high school students continue to win international science and math competitions (Educational Testing Service, 1989). Similarly, the best products of U.S. secondary and higher education in technical disciplines are considered by U.S. competitors to be a determining factor in U.S. leadership in technologies demanding a particularly high degree of individual creativity, such as in the area of applications software.[4]

Building on National Technical Assets Through Globalization

It is essential to recognize that the very areas of relative national strength for the U.S. technical enterprise have become increasingly dependent on the influx of foreign talent, technology, capital, products, and competition for their continued vitality and dynamism in recent decades. Examples are the heavy dependence of U.S. university-based research and technical education on foreign-born citizens, and the role that intense foreign competition in the auto industry has played in renewing the technological dynamism of U.S.

automakers. At the same time, as the preceding discussion suggests, the increasing technological and economic interdependence of the U.S. economy has made it increasingly difficult for the United States to capture exclusively the technical and commercial returns that flow from particular national technological endowments or strengths. Clearly the United States is neither the first nor the only industrialized nation to face the challenge of rising global technological interdependence. Nonetheless, meeting this challenge is made all the more onerous for the United States by the relative decline in the ability of U.S. citizens to capitalize on their own nation's great technological strengths.

U.S. COMPARATIVE WEAKNESSES

The intensification of global competition has underlined a number of serious weaknesses in the U.S. technical and commercial enterprise—weaknesses that generally have much more to do with the management, cultivation, and organization of the nation's human and technological resources than with their relative quality or abundance. Among the most important of these liabilities are the uneven quality and limited adaptability of the U.S. work force; the underdeveloped relationship between U.S. industry and U.S. universities; chronic underinvestment in public infrastructure and industrial plant; and the relatively limited aptitude or willingness of U.S. companies (i.e., their managerial and technical leadership) to engage in cross-function al, interfirm, or international learning across the full spectrum of technical activities.

Failures of the Educational System

There are serious problems with the supply, training, and adaptability of the U.S. work force that are in large part due to failures of the nation's educational system. Most important, public primary and secondary education in the United States is failing to prepare a technologically literate citizenry. Although the best students graduating from U.S. public educational institutions can be considered world class, the share of U.S. students exiting school with substandard educations appears to be significantly greater than in other nations such as Japan. The failure of U.S. primary and secondary education to train labor force entrants in basic skills is a powerful factor in the disappointing performance of U.S. manufacturing industry in adopting new technologies. Denied a sufficiently literate and numerate general work force, the nation's engineers and scientists are less productive than they could be. Moreover, the uneven quality of U.S. public education erodes the interest and enthusiasm of many superior students who might otherwise have chosen careers in engineering or science.

Second, **the prevailing organizational structure of U.S. manufacturing firms and its associated methods of work force organization appear to have impeded the development of a more highly skilled and versatile general work force in many industries. They appear also to have raised institutional barriers to closer collaboration among technical functions within a firm and between engineers and workers on the shop floor.** The early development and widespread application of continuous process, mass-production technologies for manufacturing have been hallmarks of U.S. industry since the late nineteenth century. Mass production demanded a high degree of functional specialization of engineering and managerial tasks and a reorganization of the work process into a series of basically unskilled, repetitive activities in service of production equipment that was highly capital-intensive and specialized. Mainly because of the large size and relative homogeneity of the U.S. market, a much greater share of U.S. manufacturers adopted mass production technology and its organizational complement than did their counterparts abroad. As a result, U.S. manufacturing industries have tended to institutionalize the separation of brain work from manual work to a greater extent than their European or Asian counterparts.

This institutional and organizational legacy, however, has put many sectors of U.S. manufacturing at a relative disadvantage to their foreign competitors in industrywide efforts to develop new work force management techniques demanded by innovations in product and process technologies of the past few decades. Hence, many U.S. manufacturers may not be as effective as their Japanese and European competitors at exploiting fully the potential skills and knowledge of shop floor workers and fostering communication and cooperation among all segments of a firm's technical and nontechnical work force (Cyert and Mowery, 1987; Piore and Sabel, 1984; Stevens, 1986).

Finally, **U.S. demand for M.S. and Ph.D. engineers promises to continue to outstrip the growth of indigenous supply during the coming decades, thereby increasing U.S. dependence on foreign sources of advanced engineering talent at a time when competition for such talent is intensifying.** The share of students graduating from U.S. engineering Ph.D. programs holding permanent or temporary visas has grown dramatically since 1970, accounting for more than 50 percent of the total throughout the 1980s (Figure 1.12). Although the precise reasons for the decline in the share accounted for by U.S. citizens are not known, a number of factors are widely believed to be responsible. These include the financial penalty to the graduate student, the perception that university engineering research is sufficiently out of touch with industry to devalue an advanced degree in the eyes of industry, and the lack of faculty encouragement to potential graduate students to pursue advanced study.[5]

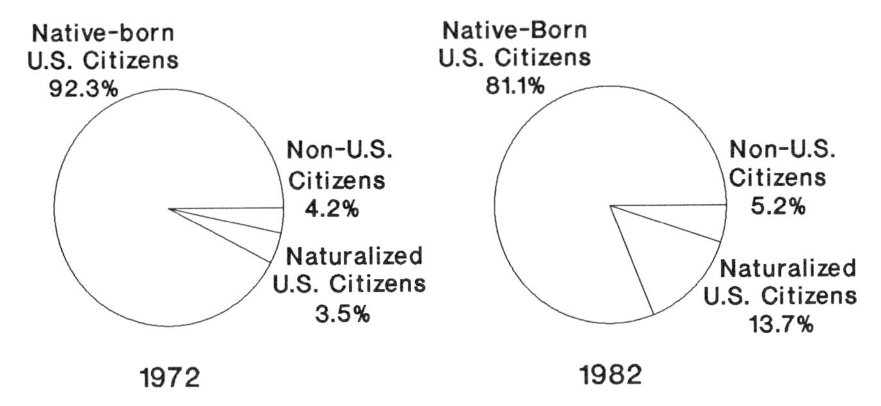

FIGURE 3.6 Composition of the U.S. science and engineering work force, by citizenship: 1972 and 1982. SOURCE: National Science Foundation (1986, p. 40).

Fortunately for the United States, foreign-born talent has bridged the gap between domestic supply and demand of advanced-degree engineers in recent years. As Figure 3.6 demonstrates, foreign-born engineers as a share of the total U.S. engineering work force more than doubled between 1972 and 1982. Moreover, as of 1982, the level of educational attainment of foreign-born engineers employed in the United States was significantly greater than that for U.S. native-born engineers (Figure 3.7).

However, if non-U.S. demand for engineering talent continues to expand and more industrializing countries follow the path of South Korea and step up their efforts to repatriate U.S.-trained engineers, it may become increasingly difficult for the United States to continue to attract the foreign talent it needs. In this context, it is also worth noting that U.S., Japanese, and other multinational corporations with subsidiaries in the newly industrialized and more advanced developing countries are currently competing for the same pool of foreign technical talent that U.S.-based firms and universities are trying to attract.

The University-Industry Mismatch

U.S. university engineering research and technical education are not sufficiently in touch with the needs of American industry.[6] Fault for this mismatch must be equally apportioned between industry and universities for not working more effectively with each other to improve engineering curricula and to reorient the research agenda of university-based engineering departments more toward the concerns of the nation's commercial engineering enterprise. The heavy reliance of the U.S. university engineering research on public money, which has been channeled predominantly toward

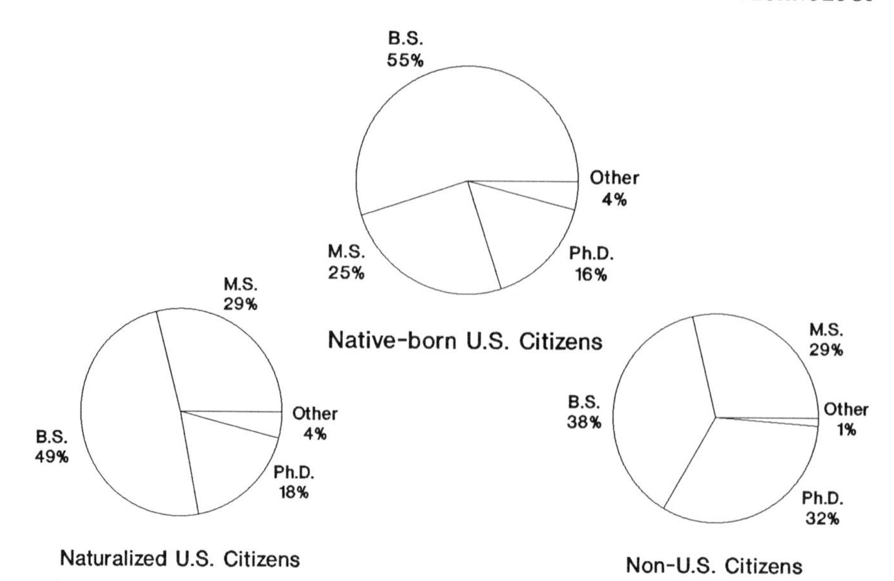

FIGURE 3.7 Educational attainment of U.S. scientists and engineers, by origin of citizenship status: 1982. SOURCE: National Science Foundation (1986, p. 40).

defense-related research since the Second World War, has also impeded greater university-industry research collaboration in a number of nondefense, commercially significant sectors.[7]

In recent years, federal budget constraints have forced U.S. universities to look to private sources for a larger share of their rapidly expanding research budgets. This has encouraged engineering and science departments to cultivate closer working relationships with industry. Recent creative initiatives sponsored by a number of state and federal agencies, such as Pennsylvania's Ben Franklin Partnership Program and NSF's Engineering Research Centers, have also helped foster greater industry-university cooperation in engineering research (National Academy of Engineering, 1989; National Governors' Association, 1988; National Research Council, 1987, 1990; Pennsylvania Department of Commerce, 1988). Despite significant progress during the past five years, however, much remains to be done to exploit the full potential of university-industry partnerships in both research and technical education.

The Eroding Economic Infrastructure

The chronically low rate of investment in the nation's economic and industrial infrastructure has undermined the productive potential of the U.S. technical enterprise and eroded the nation's industrial base.[8]

Chronically low national savings and investment rates have delayed the retraining of the U.S. work force, the modernization of U.S. industry's capital plant, and the replacement or repair of the nation's vital "social capital," that is, transportation, energy, and education infrastructures. Since the mid-1970s, U.S. real gross domestic investment as a percentage of GNP has been the lowest of the six major industrialized countries shown in Figure 3.8.[9] Moreover, while U.S. fixed investment in machinery and equipment as a percentage of GNP has grown slightly during the 1980s and remains on a par with its major West European competitors, at 8–9 percent it is still less than half that of Japan (Figure 3.9).

The causes of this running-down of the human and physical foundations of the nation's technical enterprise are complex and cannot be treated adequately within the context of this study.[10] The nation's low savings rate and its comparatively high cost of capital, together with the inability of U.S. corporate managers to combine effective short implementation cycles with long planning horizons, are among the most frequently invoked explanations. Yet, in a sense, these factors are only metaphors for a range of long-standing, deeply embedded institutional, cultural, and political impediments to investment in the foundations of wealth-generating and productivity-enhancing activities. Without a sustained effort to expand long-term investment in public infrastructure, plant modernization, and work force retraining, the

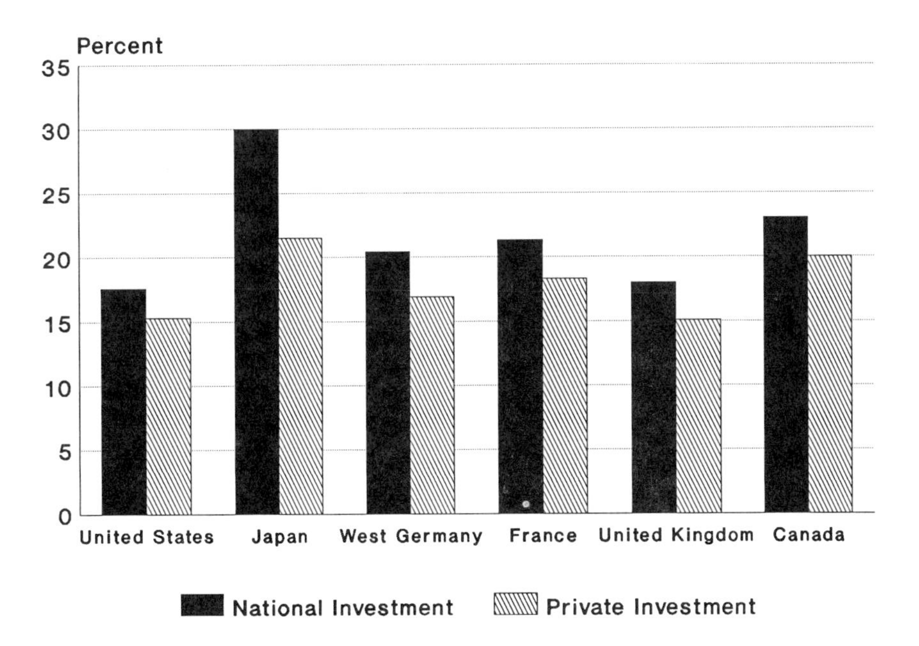

FIGURE 3.8 Gross fixed investment as a percentage of GNP, by selected countries: Average 1975–1987. SOURCE: Organization for Economic Cooperation and Development (1989).

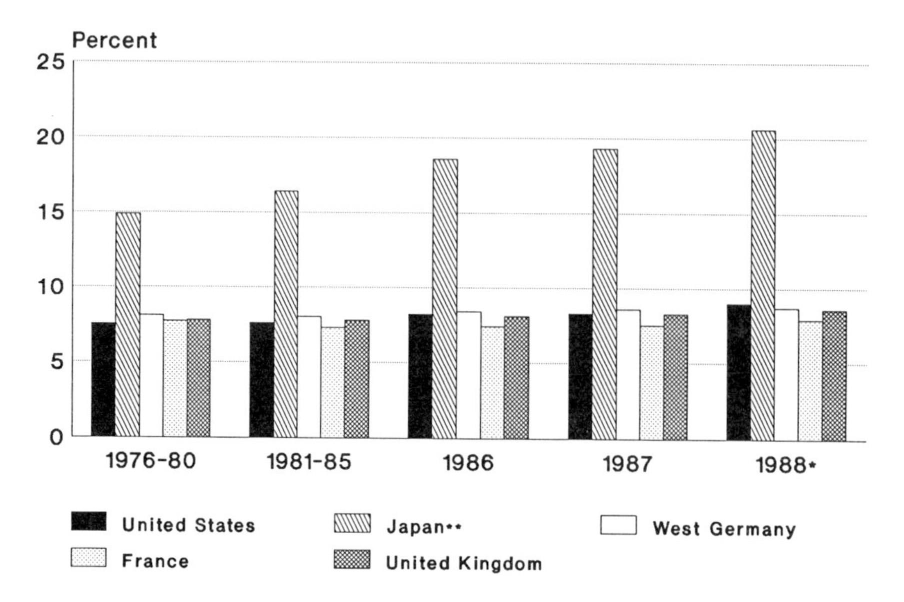

FIGURE 3.9 Fixed investment in machinery and equipment as a percentage of GNP/GDP, by selected countries: 1976–1988. *January to June. **Figures for Japan exclude public investment. SOURCE: International Monetary Fund (1989, table 17).

United States will find it increasingly difficult to leverage its vast technical capabilities for national economic growth and competitiveness.

The "Not-Invented-Here" Syndrome

The "not-invented-here" syndrome continues to inhibit the learning potential of many U.S.-based industries and thereby undercuts the nation's ability to assimilate and diffuse new technologies and engineering practices in a timely manner. This weakness manifests itself in a variety of "underdeveloped" or "lopsided" relationships, including the interaction of technical and related functions within an individual firm (the "mass production" legacy) and the links between industry and university engineering departments. It is apparent also in relations between U.S. firms and their domestic supplier base and between U.S. companies and their foreign counterparts.

Intra- and interfirm technical relationships are underdeveloped in the United States. Intense global competition and the shortening of product cycles have underlined the growing importance of combining the benefits of competition with the technological cross-fertilization, economies of innovation, and enhanced organizational learning that can result from cooperation among material and component suppliers, equipment vendors, and system integrators across a range of industries. Whereas private and public actors in Japan appear to have struck a competitive balance between the creative destruction of intense competition and mutually advantageous intra-industry technical collaboration, many of their counterparts in U.S. industry and government continue to take the benefits of technical collaboration and cross-company organizational learning for granted.[11]

The experience of the U.S. semiconductor industry illustrates pointedly how the lack of stable, long-term relationships between suppliers and system integrators within an industry can contribute significantly to a general erosion of the technological competitiveness of activities both upstream and downstream in an industry's value-added chain (see industry profile in Appendix A).

Similarly, the segregation and often faulty quality of communication between various technical, managerial, marketing, and other functions in firms continues to deny many U.S. companies the economies of functional integration and cross-functional learning so critical to rapid development and commercialization of innovations (Gomory, 1989).

Much of U.S. industry remains unreceptive to, ignorant of, or incapable of exploiting foreign technical advances and foreign engineering and management practices. It is commonly assumed that U.S. transnational corporations are sufficiently attuned to global markets to track technological developments in other countries that might affect the competitiveness of firms in their industries. Yet the experiences of several large U.S. corporations illustrate that knowledge of foreign technical capabilities is not always accompanied by the wisdom or ability to act on that knowledge. For example, the U.S. materials industry's failure to appreciate the significance of overseas advances in wafer technology during the early 1980s, despite warnings by some of its U.S. customers, soon cost it world leadership in that technology and has ultimately contributed to a rapidly growing dependence of U.S. semiconductor manufacturers on a very limited number of foreign suppliers of wafer technology and product.

Even when the tracking of global technology and know-how is performed well by U.S. multinational corporations, they frequently fail to transfer the acquired technology, engineering practices, or managerial techniques to their own plants, supplier base, or downstream customers in the United States (Mansfield, 1988).

NOTES

1. Some industry experts argue that in many sectors, such as microelectronics, Japanese consumers are more demanding of quality, performance, design, and service than American.
2. In 1987 the pool of capital managed by U.S. venture capital enterprises totaled $29 billion, eight times the amount available in 1978. This large venture capital reservoir

played a critical role in the phenomenal growth in a number of small high-tech businesses during the 1980s. From 1981 to 1986 alone, the United States experienced a net gain of approximately 30,000 high-tech firms and a corresponding jump in the small high-tech business work force from around 89,000 to 158,000 employees. See National Science Board (1989, pp. 141-145, 363-370).

3. The extremely high mobility of the U.S. technical work force may be a "mixed blessing," to the extent that it discourages U.S. employers from investing in the continuing education and training of their employees. See National Academy of Engineering (1988).
4. There is some debate whether a trade-off exists between creativity and discipline, whether the U.S. educational system tends to cultivate creativity at the expense of self-discipline, good work habits, and attention to the details of execution, which may be as important for competitiveness as originality.
5. It is noteworthy that federal support for graduate fellowships began to decline in the early 1970s, coinciding with the decline in U.S.-born enrollments in engineering doctoral programs.
6. Indeed, in comparison with university-industry relationships in other countries, U.S. universities enjoy relatively close ties with American industry. However, given the fact that U.S. universities play a much more central role in the U.S. total research enterprise than their counterparts do in Asia or Europe, it is that much more critical to the United States that its universities and industry work closely together.
7. U.S. universities frequently accuse U.S. industry of assuming a delinquent "parishioner" attitude toward university research—financial contributions with little, if any, human capital support or follow-up. At the same time, university-based researchers rebut arguments regarding the industrial relevance of their work by pointing to the growing interest of foreign corporations in areas of U.S. university-based research that have been neglected by U.S. companies; for example, civil engineering and construction research.

   U.S. industry, on the other hand, decries the academic research community's general disdain for industry-specific research problems. Moreover, by their own admission, U.S. corporations rely much more heavily on gaining access to university research capabilities through the hiring of faculty members as consultants and graduate students as engineers than their foreign counterparts. Moreover, since these indirect "human" transactions tend not to appear on university balance sheets, other more direct forms of support for university research—those most often practiced by foreign companies and governments, such as financial, material, and institutional support—may overstate the disparity of U.S. and foreign corporate interest in U.S. university research.
8. In a recent article in the *New England Economic Review,* Munnel (1990) argues that the abrupt drop in productivity growth in all the OECD countries after 1975, despite a continuing high level of R&D investment, could be attributed to the dramatic fall off in public infrastructure investment beginning in the mid-1970s. This suggests that public infrastructure is as important as the knowledge stock in stimulating productivity growth.
9. Available comparative data on investment rates and the cost of capital in different countries do not account for significant differences in accounting procedures among countries, and are therefore believed to overstate international differences. Recent studies have argued that national variations in the way capital is depreciated and the categorization of government expenditure account for at least part of the wide gap between U.S. and Japanese savings and investment rates. See, for example, Hayashi (1989); McCauley and Zimmer (1989).

10. The National Academy of Engineering Committee on Time Horizons and Technology Investments has explored several important aspects of this topic. The committee's findings are expected to be published in the spring of 1991.
11. In his most recent work, Michael Porter (1990) offers a valuable warning against the anticompetitive, or collusive, potential of corporate alliances and consortia, and presents a strong case for ensuring that competition is not compromised by such initiatives.

    For all their success at combining competition with cooperation, the Japanese have been repeatedly criticized for the excessively or "collusively" tight linkages between Japanese firms within certain industries that effectively prohibit foreign companies from participating in all-Japanese value-added chains. See, for example, Prestowitz (1988).

REFERENCES

Cyert, Richard M., and David C. Mowery, eds. 1987. Technology and Employment: Innovation and Growth in the U.S. Economy. Washington, D.C.: National Academy Press.

Educational Testing Service. 1989. A world of differences: An international assessment of mathematics and science. Report No. 19-CAEP-01. Princeton, N.J.: Educational Testing Service.

Gomory, Ralph. 1989. From the 'Ladder of Science' to the product development cycle. Harvard Business Review (Nov-Dec):415-421.

Hayashi, Fumio. 1989. Is Japan's saving rate high? Federal Reserve Bank of Minneapolis Quarterly Review (13)2:3-9.

International Monetary Fund. 1989. World Economic Outlook, Washington, D.C.: IMF.

Mansfield, Edwin. 1988. The speed and cost of industrial innovation in Japan and the United States: External vs. internal technology. Management Science (34)10:1157-1168.

McCauley, Robert N., and Steven Zimmer. 1989. Explaining international differences in the cost of capital. Federal Reserve Bank of New York Quarterly Review (Summer):7-28.

Munnell, Alicia H. 1990. Why has productivity growth declined? Productivity and public investment. New England Economic Review. (Jan-Feb):3-22.

National Academy of Engineering. 1988. Focus on the Future: A National Action Plan for Career-Long Education for Engineers. Washington, D.C.: National Academy Press.

National Academy of Engineering. 1989. Assessment of the National Science Foundation's Engineering Research Centers Program. Prepared for the National Science Foundation. Washington, D.C.: National Academy of Engineering.

National Governors' Association. 1988. State-Supported SBIR Programs and Related State Technology Programs. Center for Policy Research and Analysis. Prepared by Marianne K. Clarke for U.S. Small Business Administration, Washington, D.C. February.

National Research Council. 1987. The ERCs: Leaders in Change. Commission on Engineering and Technical Systems. Washington, D.C.: National Academy Press.

National Research Council. 1988. Foreign and Foreign-Born Engineers in the United States. Office of Scientific and Engineering Personnel. Washington, D.C.: National Academy Press.

National Research Council. 1990. Ohio's Thomas Edison Centers: A 1990 Review. Commission on Engineering and Technical Systems. Washington, D.C.: National Academy Press.

National Science Board. 1989. Science and Technology Indicators—1989. Washington, D.C.: U.S. Government Printing Office.

National Science Foundation. 1988. International Science and Technology Update: 1988.

Research and Development (R&D) Expenditures. Special Report NSF 89-307, Detailed Statistical Tables. Washington, D.C.: National Science Foundation.

National Science Foundation. 1986. Foreign Citizens in U.S. Science and Engineering: History, Status, and Outlook. Special Report NSF 86-305, Surveys of Science Resources Series. Washington, D.C.: National Science Foundation.

Organization for Economic Cooperation and Development. 1989. National Accounts (Volumes 1 and 2). Paris: OECD.

Pennsylvania Department of Commerce. 1988. Ben Franklin Partnership: Challenge Grant Program for Technological Innovation—Five Year Report. Board of the Ben Franklin Partnership Fund. Harrisburg, Pa.: Pennsylvania Department of Commerce.

Piore, Michael, and Charles Sabel. 1984. The Second Industrial Divide. New York: Basic Books.

Porter, Michael E. 1990. The Competitive Advantage of Nations. New York: The Free Press.

Prestowitz, Clyde V., Jr. 1988. Trading Places. New York: Basic Books.

Stevens, Barrie. 1986. Training for Technological Change. Centre for European Policy Studies, CEPS Papers, no. 31. Brussels: Centre for European Policy Studies.

U.S. Bureau of the Census. 1988. Recent Data on Scientists and Engineers in Industrialized Countries. Report prepared for National Science Foundation, Division of Science Resources Studies. Center for International Research. February.

U.S. Department of Commerce. 1989a. U.S. Industrial Outlook—1989. Washington, D.C.: U.S. Government Printing Office.

U.S. Department of Commerce. 1989b. Staff Report—Direct Investment Update: Trends in International Direct Investment. International Trade Administration, Office of Trade and Investment Analysis. Washington, D.C. September.

U.S. General Accounting Office. 1988a. R&D Funding: Foreign Sponsorship of U.S. University Research. Report to the Honorable Lloyd Bentsen, U.S. Senate. March. Gaithersburg, Md.: U.S. General Accounting Office.

U.S. General Accounting Office. 1988b. Technology Transfer: U.S. and Foreign Participation in R&D at Federal Laboratories. Report to the Honorable Lloyd Bentsen, U.S. Senate. August. Gaithersburg, Md.: U.S. General Accounting Office.

4

Capturing Benefits of Global Technical Advance: Policy Implications

Most competitiveness studies of the past decade have focused on what U.S. industry, government, and universities need to do within the United States to improve the productivity, quality consciousness, and cost effectiveness of U.S. firms, thereby enabling them to compete more effectively with foreign firms at home and abroad. Three major implicit assumptions underlie most of these studies: (1) the level of transnational technological interdependence is relatively low, that is, from the U.S. perspective, cross-border technology flows remain predominantly unidirectional out of the United States; (2) it is possible to distinguish one nation's technology, companies, products, and investments relatively neatly from those of its trading partners; and (3) the scope and impact of domestic and international policies remain relatively well delineated and discrete.

The domestic policy recommendations of these studies have generally focused on ways to encourage more efficient and effective management of U.S. productive resources—human capital, technology, raw materials, physical plant, capital, etc. International policy recommendations have tended to focus on issues regarding trade, intellectual property rights, and international standards while devoting little attention to issues raised by transnational technology flows or foreign direct investment.

During the past decade, however, the rapid advance of international technological convergence and interdependence have recast the "competitiveness" challenge to U.S. policymakers and the nation's technical community. Ten years ago it may have still made sense for U.S. lawmakers to focus attention almost exclusively on unilateral initiatives to strengthen the national technological base and open foreign markets to U.S.-engineered products and services. These measures, however, do not deal adequately with the high volume of intrafirm trade, foreign direct investment (both inward and outward), and extensive global technical networks that define the current international economic order. Nor is the traditional competitiveness framework well equipped to deal with the changing character of competition among nations that has accompanied the globalization of technology and industry.

GLOBALIZATION OF ADVANCED TECHNICAL ACTIVITIES

As the preceding chapters have shown, the globalization of advanced technical activities is a well-established trend driven by a number of powerful economic, technological, and political imperatives. Indeed, there is considerable variation in the scope and character of globalization among industries. The nature of each industry's technology, its product and production processes, the structure of its value-added chain, and the scale and scope of its market all play important roles in explaining interindustry variations. Likewise, the extent to which firms within a given industry have globalized their technical activities often varies significantly according to each firm's tenure in international markets, the size and openness of its home market, and the peculiarities of its historical development within a given national economy (see Appendix A). Nevertheless, the momentum and broad scope of the globalization trend are apparent.

Today technological capabilities are much more widely distributed throughout the globe than they were 10 or 20 years ago. In this new multipolar technological order, the activities of multinational corporations have contributed to the emergence of a transnational technology base in a growing number of industries. As a result, it is becoming increasingly difficult to distinguish one firm's technology from another's or one nation's technology base from another's.

The economic and technological drivers of globalization are intimately linked. The diffusion of technological advance across national borders is rapid and accelerating. There are now multiple sources of technological advance dispersed throughout the globe. With the intensification of global competition, technology lifecycles are shortening in many sectors. This makes it increasingly difficult for firms to recoup investments in research and development, the costs of which have been spiraling upward in many industries. In this new environment, technological self-sufficiency is a prohibitively expensive luxury that almost no firm can afford. Moreover, technological breakthroughs offer small prospect for enduring competitive advantage if they are not complemented by vigilance and vigor in the pursuit of continuous product and process improvement and the reduction of product cycle time. To compete effectively in a global economy, corporations must be able to draw on the broadest possible range of technical capabilities worldwide, exploiting the special competencies and demands of many different national markets through foreign direct investment, transnational corporate alliances, and trade. In short, corporate survival in an era of intense global competition and internationally dispersed technical capabilities demands global corporate strategies and conduct.

The political imperatives of globalization are equally compelling. Many countries, developed as well as developing, are not willing to accept the current international division of labor. It is only natural that nations should aspire to higher levels of technical competence and the higher productivity and wealth-generating capability that are associated with high-tech or technology-intensive industries. In consequence, many countries have developed policies designed to persuade or compel multinational corporations to locate a greater share of their production and other advanced technical activities within their borders. For example, local content requirements and other nontariff barriers to trade make it virtually impossible for a firm to penetrate certain foreign markets except through foreign direct investment, joint ventures, or other technology-transfer or production-sharing arrangements.

THE CHANGING CHARACTER OF COMPETITION AMONG NATIONS

The competition for economic and technological advantage among nations has intensified, not lessened, with the globalization of industry and corresponding growth of international economic interdependence. Governments have long intervened in their domestic economies in an effort to increase the productivity and international competitiveness of firms operating, if not originating, within their borders. However, as more and more countries have come to recognize the importance of technological advance for economic growth and competitiveness, the quest for economic advantage among governments has come to focus more intensively on creating a domestic environment that is conducive to the development, adoption, adaptation, and diffusion of advanced technology by private companies for commercial advantage.

In this new competition, governments have, for the most part, eschewed the traditional, more transparent instruments of economic statecraft, such as tariffs and quotas. However, they continue to compete fiercely, if indirectly, to attract the resources and wealth-generating potential of private corporations through more subtle "domestic" policy mechanisms such as subsidies, tax credits, deregulation, domestic content legislation, public procurement, or other public policies designed to strengthen their domestic economic or technical infrastructures.

It is important to recognize, however, that in policy areas where no internationally accepted rules apply, competition for economic advantage among nations is by nature "unfair." Some nations are better at absorbing the technology of others or are more adept at closing their markets to foreign competition. National "unfair" advantages may stem from specific policies or from "structural" differences, such as legal, institutional, or cultural characteristics. In any event, these national differences in industrial organization and political economy raise issues of equity—reciprocal access to national markets and technical resources, or the problem of "free-riders" on the global research base—that are bound to become more and more politically contentious as the globalization trend gathers momentum.

To deal effectively with the domestic and international political friction generated by the global integration of national or regional technology markets, national governments are being called upon to negotiate sovereignty in areas of public policy traditionally viewed as exclusively matters of domestic concern. The changing corporate strategies of multinational corporations and the rapid growth of international economic and technical interdependence underline the fact that the neat dichotomy of domestic and foreign policies—the distinction between policy areas that are "legislated" domestically and those that are "negotiated" internationally—is breaking down.

Domestic regulations regarding competition, health, safety, and the environment, fiscal and monetary policy, and a broad spectrum of science and technology policies, all are forcing their way onto the international negotiating table despite the protestations of national governments. International variations in the regulation and enforcement of competition or antitrust policy, for example, affect the relative accessibility of national markets to foreign trade or investment. The Japanese *keiretsu* (cartel-like industrial groups), which are considered major barriers to foreign corporate entry into the Japanese home market and sources of potentially anticompetitive oligopoly power in global markets, clearly would not be tolerated under U.S. or European antitrust law. On the other hand, the regulatory maze that foreign firms in search of U.S. acquisitions are compelled to negotiate in the United States in industries such as insurance is also an impediment to cross-border investment and corporate activity. Similarly, international variations in health and safety policies, tax policies, or environmental standards influence cross-border flows of goods, capital, technology, and people in complex, yet significant, ways. Indeed, the expanding scope of multilateral negotiations within the European Community and the General Agreement on Tariffs and Trade (GATT), and bilateral negotiations such as the U.S.-Japan Structural Impediments Initiative, all reflect the fact that "domestic" issues have become virtually inseparable from the established international policy debates concerning international trade and investment.

IMPLICATIONS FOR THE UNITED STATES

The rapid growth of international technological and economic interdependence in recent years poses a number of major opportunities and challenges to the United States. As the preceding chapters have suggested, in many respects the United States is well positioned to benefit disproportionately from the positive-sum dynamic of globalization. The sheer size, technical sophistication, and relative openness of the U.S. market, the superiority of the U.S. basic research enterprise and national information technology network, the country's large pool of technical talent, and the overall attractiveness of the American sociopolitical system are all tremendous sources of national strength in the emerging global economy.

Nevertheless, as we have seen, the ability of the United States to take full advantage of and capture a fair share of the benefits of the emerging global technical enterprise is challenged by a number of factors and forces originating both within and beyond the nation's borders. For more than half a century, the United States has been the world's dominant and most autonomous technological power. During this period, the competitive strength of the U.S. industrial base was generally understood to be the product of the superior R&D capabilities and the advanced product and process technology of U.S.-owned corporations. Technical leadership, along with abundant capital and extensive experience with large-scale distribution systems, was thought to compensate for relatively high U.S. labor costs.

The lessons learned by the United States during its era of unrivaled technical supremacy have become singularly dysfunctional as the technology and income gaps between the United States and its trading partners have closed. To be sure, the United States remains the world's premier source of science and technology, excelling particularly in research and the application of sophisticated information technology to the integration of complex systems. Furthermore, the U.S. economy is arguably still less dependent on foreign technology than any other industrialized country. However, since the mid-1970s the rapid penetration of the U.S. domestic market through foreign direct investment and trade, the growing globalization of the advanced technical activities of U.S. multinational corporations, and the concurrent loss of world market share and relative technical capability by numerous U.S.-based industries have underscored the limitations and vulnerabilities of the U.S. technical enterprise in its present state.

The last two decades have shown that abundant scientific and technical resources and production of new knowledge or "cutting edge" technology by themselves do not add up to commercial success. For all of its pronounced advantage in research and invention, the U.S. technical enterprise as a whole has demonstrated considerably less aptitude than its Japanese counterpart in the rapid adoption, adaptation, and diffusion of technology for commercial gain. Furthermore, it is apparent that U.S. citizens and U.S.-based companies are, for the most part, relatively poorly equipped organizationally, financially, culturally, and educationally to take advantage of new technology, domestic or foreign, originating outside of their own department, division, or firm.

Clearly, not all of the challenges facing the United States in this regard are domestic in origin. The lack of international consensus or rules with regard to high-tech trade, foreign direct investment, antitrust policy, or the host of increasingly international—formerly "domestic"—policy issues makes it increasingly difficult for U.S. and foreign legislators to resist the protectionist backlash that accompanies most structural adjustment and economic change.

POLICY DIRECTIONS

One major conclusion emerges from this study. The globalization of technology and industry has radically altered the nature of the industrial competitiveness debate. **In this new global environment, the highest priority for strengthening the technological foundations and thereby the long-term wealth-generating capacity of the U.S. economy must be to make the United States a more attractive and advantageous place for individuals, companies, and other institutional entities, regardless of national origin, to conduct the full complement of technical activities critical to the nation's long-term prosperity and security. To accomplish this, the United States must develop the necessary human, financial, physical, regulatory, and institutional infrastructures to compare more advantageously with other nations in attracting the technical, managerial, and financial resources of globally active corporations or individuals.**

All sectors of U.S. society—industry, government, and both basic and higher education—have important roles to play in this effort. The committee has focused primarily on public policy implications, but it does not believe that public policies are the only or even the most important determinants of national or corporate technical strength and competitiveness. Rather, the study's public policy focus has been shaped by the fact that the public sector is groping to formulate and implement a national agenda that can address the imperatives of a highly integrated global economic and technological order.

The government must take action on many fronts to strengthen the foundations of the U.S. technical enterprise—the nation's work force, its social capital (i.e., educational system and public infrastructure), as well as its fiscal and regulatory environment. **Above all, state and federal policymakers**

must work together with corporate and academic leaders to develop a broad national consensus regarding the need to improve technology development, adoption, adaptation, and diffusion throughout the U.S. industrial economy. This consensus, in concert with other national policies, can provide the necessary impetus, coherence, and operational guidelines for the many diverse private and public policy actions required to meet the challenges of globalization.

Given the relative strengths and weaknesses of the U.S. technical enterprise and inherent challenges and opportunities of the emerging global technical enterprise, the following domestic and international policy directions represent, in the judgment of the committee, essential elements of a broad national technology strategy for the United States.

Domestic Policy Directions

It is imperative that the United States continue to build on the comparative strengths of the nation's technical enterprise—its research capabilities; its system of advanced technical education; its large pool of elite technical talent; and its extensive, sophisticated information technology infrastructure. These comparative advantages find expression in continuing U.S. commercial leadership in highly science-intensive industries or industries in the infancy of their technology life cycle. Largely as a result of their extensive research activities, U.S. universities are able to provide critical high-quality human and intellectual inputs to many high-technology industries, most of it to U.S.-owned and U.S.-based firms. Because of its strength, the U.S. research enterprise has served as a magnet to foreign talent and foreign investment, both of which have contributed substantially to U.S. economic growth and prosperity in recent decades. Finally, the apparent trend in many industries toward a technological future that is increasingly science-based suggests that proximity to U.S.-based research capabilities and this country's superior information network will become increasingly important for domestic- and foreign-owned corporations alike.

In this context it is cause for concern that the efforts of U.S.-based industry in basic and long-range applied research appear to have faltered in recent years. From 1974 to 1984, U.S. industry spending on basic and applied research grew at average annual rates of 23 and 27 percent, respectively. Yet between 1984 and 1989 the average annual rate of growth of corporate spending on basic research had fallen to 4 percent, while that for applied research dropped to 6 percent. A prolonged decline or stagnation in basic and long-term applied research in the private sector is almost certain to have negative long-term consequences for the technical strength and competitiveness of U.S.-based companies. Among other things, without a vital and expanding R&D effort, U.S.-based companies will find it increasingly difficult to exploit the R&D efforts and capabilities of their foreign competitors, let alone enter into mutually beneficial technical alliances with them. **Therefore, the committee believes it is essential that the federal government continue to help foster increased basic and long-term applied research by the private sector.**

Policy Initiatives to Enhance Technology Adoption, Adaptation, and Diffusion

Nevertheless, the past two decades have also demonstrated for a broad range of manufacturing and service industries that as new knowledge flows more freely across national borders, the ability of a nation or a firm to exploit research results for commercial advantage depends increasingly on mastery of downstream technical activities associated with product and process development and production more generally. This trend is particularly troublesome for the U.S. technical enterprise, whose comparative weakness is most pronounced in the areas of product and process technology development, adoption, adaptation, and diffusion—weaknesses that find clear expression in the performance of many U.S. manufacturing industries, particularly where the pace and direction of technological advance are less dependent on basic research or less prone to revolutionary technical breakthroughs.

The relative decline in the ability of U.S. citizens to derive commercial benefits from an increasingly cosmopolitan technology base underlines the need for U.S. educators, industrialists, and policymakers to direct greater attention and resources toward the "relearning" of the less prestigious, yet equally vital, activities of technology adoption and adaptation. These are competencies, after all, that are closely associated with the production of goods and services in which the United States excelled from the late 1800s well into the mid-1900s. By highlighting these national vulnerabilities, the process of globalization has lent greater urgency to oft-repeated calls for public and private initiatives to bolster the nation's production engineering capabilities and its overall manufacturing base.

Current federal science and technology policies are targeted toward basic research and "mission-oriented" technology development, mainly in the areas of national defense, public health, and space exploration—areas generally accepted as primary responsibilities of the federal government. While building on the comparative strengths of the nation's technical enterprise, this policy orientation neglects pressing national vulnerabilities that have less to do with an inability to create new technology than with a failure to adopt and adapt existing technology effectively for commercial benefit. **Therefore, the committee recommends that policymakers expand support for diverse initiatives at the federal, regional, and state levels to**

enhance the adoption, adaptation, and diffusion of industrial technology and related know-how. Recent U.S. experience has demonstrated that low-cost, pragmatic state and federal policy initiatives can support private-sector progress in these areas. The National Science Foundation's Engineering Research Centers, Ohio's Thomas Edison Program, Pennsylvania's Ben Franklin Partnership Program, the Southern Technology Council, or the Industrial Technology Institute are promising means for providing public support for a diverse set of initiatives and selectively broadening the application of those that prove most successful (see National Academy of Engineering, 1990; National Governors' Association, 1988; National Research Council, 1990; Pennsylvania Department of Commerce, 1988).

A New Approach to Generic Technology Development

The intensification of international technological competition and interdependence underlines the need for the United States to develop a broader approach to the development and diffusion of commercially significant generic technologies. Such technologies involve concepts of design, fabrication, and quality control applicable to a class of products for which (a) the anticipated returns from development and commercialization cannot justify the expense and risk of investment by single firms or joint ventures, and (b) the returns to the economy and society as a whole warrant investment by the federal government. In addition, there may be areas in which national military strategic considerations make unacceptable the loss of U.S. technology position or market share.

Promotion of commercially significant generic technologies need not require major investments in research and development programs. Indeed, obstacles to the diffusion of such technologies may be more important than any obstacle to their development. To be sure, significant public and private investment may be required in certain cases, as in the development of a new generation of semiconductors, when the cost of technological advance is so high, the time scale of technology development is so long, and the ability of any one firm to benefit from such large investments is so low or unpredictable that no firm is willing to take the risk. For other generic technologies, however, development costs may not be high—or the technology may already be available—yet there may be serious economic, regulatory, or cultural obstacles to the adoption, adaptation, and diffusion of the technology either within or across industries. For example, "total quality control" methods, computer-aided design, advanced construction techniques, and just-in-time production systems are all generic technologies that might fall into this category.

Given the current extent of global technological interdependence and the relative strengths and weaknesses of the U.S. technical enterprise, a new, more inclusive approach to the development and diffusion of commercially significant generic technologies is needed. To begin with, such an approach should complement the development or relearning of specific technologies and technical competencies by U.S.-based firms with much more aggressive and methodical efforts at tracking and exploiting relevant foreign technical and managerial capabilities. Admittedly, the best way for the nation to assimilate new technologies or learn new ways to manage product and process technology development more efficiently and effectively is for corporations based within its borders to scale the learning curves associated with the relevant technologies themselves. In other words, these are technical competencies better acquired actively than passively through the acquisition of "off-the-shelf" product or process technologies. This does not, in itself, argue against the participation of foreign-owned firms in the national technology regeneration process. On the contrary, given the high level of technical and managerial competence demonstrated by foreign firms in any number of high-, medium- and low-tech industries, it is all the more urgent that the U.S.-based corporations improve their ability to identify and draw on significant technical and managerial innovations, whatever their origin, in a more timely manner. This can be accomplished through arm's-length transactions—technology licensing, technology tracking, or transnational managerial and technical personnel exchanges—or by encouraging foreign firms to participate in consortia or other collaborative arrangements.

Second, a more inclusive approach to generic technology development should recognize that the indirect benefits of public investments in generic technology development are frequently as beneficial to U.S. national interests as the specific technical processes or products that might result from such ventures. Potential by-products of collaborative public-private generic technology programs include the cultivation of local or regional technical networks; the resulting diffusion of "best practice" technical, managerial, and organizational capabilities; and the enhanced intercorporate organizational learning that enables participating firms to translate related technologies into commercial products rapidly and effectively.

In this context, public policies in support of generic technology should be more attentive to the broader economic and regulatory factors that might amplify or diminish the direct and indirect benefits of generic technology programs. Such factors include the quality and quantity of human, financial, and physical capital; the health of the local, regional, or national supplier and customer bases; and the nature and extent of competition within the affected industry sectors. In other words, targeted policy initiatives should be better coordinated with and balanced by efforts that ensure the availability of the complementary resources necessary for U.S.-based corporations to profit from the output of more focused generic technology efforts. These efforts would not prevent the benefits of publicly funded precompetitive research from flowing to foreign-based corporations; however, they would help ensure that U.S.-based firms are better equipped to capture their fair share of the commercial returns.

There is at present considerable debate regarding the proper government role in support of generic technologies. In the opinion of the committee, the primary role of government should be as convener and catalyst of activities undertaken in the private sector. In some cases this may involve federal matching of a significant amount of private funding. However, in most instances the government should be prepared to serve as the "pathfinder," providing more indirect fiscal or regulatory support to private-sector participants. Ultimately any effort to provide government support for the development and diffusion of generic technology in the United States will depend on the credibility of the public and private institutional mechanisms designated to assess and identify those technologies most in need of attention and to chart an appropriate policy response. The committee notes that there have been several attempts by federal agencies to identify "critical" technologies in recent months, most notably by the departments of Commerce (1990) and Defense (1990). The mixed reception of these efforts in the U.S. policy community, however, emphasizes that institutions that perform this function should be perceived as technically expert, responsive to the interests of all U.S. citizens—consumers, producers, and suppliers—and predisposed to operate in a manner consistent with emerging global economic and technological realities.

In summary, U.S. public policy should acknowledge the need for a stronger public role in support of generic technological capabilities for the benefit of the nation, and establish credible mechanisms for translating this commitment in principle into specific actions.

The Issue of National Treatment

The rapid increase in foreign direct investment in U.S. technology-intensive manufacturing and service sectors and the extensive involvement of U.S.-based firms in transnational technical alliances has blurred the distinction between "U.S. firms" and "foreign firms" to the point that it has become nearly impossible for the government to find purely "indigenous" corporate partners with which to pursue national industrial, technological, and economic objectives. In light of the current contribution of foreign corporations to U.S. economic growth and technological strength and the transnational character of many U.S.-owned companies' technical activities, is seems only appropriate that public sector assistance to, or collaboration with, private corporations (domestic or foreign) in pursuit of national objectives should be governed by common standards for the corporate role in the U.S. economy.

It is perfectly reasonable and correct for the federal government to expect and require all corporations that benefit from access to the U.S. market or participate in publicly supported technology initiatives to demonstrate good corporate citizenship, that is, to abide by their host government's laws and regulations. Furthermore, it is entirely appropriate that policymakers charged with advancing the interests of all U.S. citizens should develop criteria consistent with that charge regarding corporate participation in any venture involving public funds or legal exemptions. However, in a global economy with globally active corporations, corporate nationality is a poor measure of a firm's real or potential contribution to the U.S. economy.

This is not to say that the issues of corporate ownership and control have become totally irrelevant to the pursuit of national economic and technology interests. There may be circumstances in which the U.S. government should discriminate against foreign-owned firms temporarily to achieve reciprocal equitable "national treatment" of U.S. companies doing business overseas or to safeguard national security. Nonetheless, recent growth in foreign technical capabilities and international technological interdependence suggest that as U.S. lawmakers assess the relative costs and benefits of discriminatory policies, they should attach greater weight to the many benefits U.S. citizens derive from foreign participation in the domestic market through increased employment, real wages, technology transfer, and competition. **In summary, public policy initiatives that seek to strengthen the nation's commercial technology and industry base should be guided by the extent to which a corporation genuinely contributes to the national economy. With rare exceptions such policies should not discriminate among corporations on the basis of nationality of ownership or incorporation, provided there is sufficient reciprocity in the large.**

Improving the Nation's Work Force and Economic Infrastructure

The globalization of technology and the intensification of competition among firms and nations impart a new sense of urgency to long-standing recommendations regarding state and federal government support for modernizing and improving the quality of the nation's human and social capital (Council on Competitiveness, 1988; National Academy of Engineering, 1988a, 1988b; President's Commission on Industrial Competitiveness, 1985). Clearly, new or more technology by itself will not generate the wealth or productivity increases necessary to increase the standard of living of U.S. citizens and strengthen U.S. national competitiveness. These objectives demand that the United States devote greater attention to the social and human capital that supports the technological capabilities and commercial vitality of corporations based or operating within U.S. national borders.[1] Government has a critical role to play both directly, through public investment in the nation's educational system and physical infrastructure, and indirectly, through leadership in encouraging industry and universities to become more involved in efforts to improve the quality of the U.S. work force. Similarly, government plays an important role in creating a fiscal and regulatory environment that will encourage private industry to make investments in plant, equipment, and organizational learning that will enable it to adopt, adapt, and create value from technological advances. **Therefore, the committee urges state and federal governments to redouble their efforts to modernize and strengthen the nation's work force and public infrastructure, and to encourage private industry to modernize its plant and equipment.**

Technical Competence in Government

The development and commercialization of technology are not a discrete policy issue, but an integral part of many broader areas of domestic and foreign policy. Until recently, there has been insufficient appreciation across agencies of the implications that various policy initiatives hold for science and technology. There has been even less communication and cooperation among those responsible for formulating and implementing domestic and foreign policies that bear on the health of the nation's commercial technology base. This situation argues for expanding agencies' recruitment of technically competent personnel to formulate and implement domestic and international economic policy; it also points up the need for greater organizational focus at the national level on the policies affecting commercial development and application of technology.

The committee notes with guarded optimism the positive steps taken by the current administration to provide more organizational focus through the President's Science and Technology Adviser, recently elevated to the position of Assistant to the President, the President's Council of Advisers on Science and Technology, the Office of Science and Technology Policy, the newly created Office of Technology Policy in the Department of Commerce, and Commerce's National Institute of Standards and Technology. These bodies clearly have the potential for improving intragovernmental communication and coordination across a range of domestic and international policy areas related to technology and economics. Ultimately, it is of secondary importance whether the necessary organizational focus is located in a single independent agency (existing or to be created) or finds expression in more institutionalized interaction among the many agencies and committees that currently influence the nation's technology base. What is critical is that those seeking to develop greater organizational focus acknowledge the growing synergies between what have traditionally been viewed as discrete policy areas.

In summary, the committee recommends that the federal government devote greater attention to the technological dimensions of international trade, investment, competition, and other critical issues not traditionally associated with science and technology concerns. To this end, government should seek to cultivate greater technical expertise in agencies responsible for domestic and international economic policy and to improve interagency communication and coordination regarding science and technology issues.

International Policy Directions

The increasingly global character of corporate technical activities has made it essential that policies aimed at developing and better managing the nation's technical endowments be outward looking—consistent with an international policy framework that fosters and structures technological competition, cooperation, and exchange among nations and firms. Ultimately, the nation's ability to capture a fair share of the benefits of the global technical enterprise will depend primarily on the extent to which private corporations operating within its borders seize the opportunities presented by the emerging global technology base. Their success or failure, however, will be conditioned by the extent to which U.S. policymakers recognize the interdependence of domestic and international policies that influence technology development, diffusion, and commercialization.

In foreign relations, there are a number of things the United States can do to complement domestic efforts, promote more reciprocal technical exchange, and attenuate tendencies toward technology-based protectionism. There is an obvious need for continued efforts to liberalize world trade as well as greater public and private involvement in the international standards-setting process, and in the quest for a more effective international intellectual property rights regime. Yet, these high-profile concerns are distracting policymakers from equally important issues raised by the rapid growth of foreign direct investment and transnational corporate alliances and technical networks over the past decade. From the perspective of the U.S. technical enterprise, one of the important challenges to U.S. foreign economic policy relates to national disparities in the treatment of foreign direct investment and competition policy.

Mutual Obligations of Multinational Corporations and Governments

In an effort to improve the nation's trade balance, and to respond more forcefully to a lack of reciprocity overseas, some recent U.S. legislation raises issues related to the free flow of foreign direct investment and to the treatment of the subsidiaries of foreign-owned corporations.[2] The rapidly increasing foreign penetration of the U.S. economy in the past two decades has stimulated concern among many segments of the American electorate. Furthermore, the discriminatory treatment of U.S.-owned corporations appears to be a fact of life in Japan and to be increasing in Western Europe as the countries of the European Community search for ways to come to terms with intensifying global competition and the consequences of EC 1992. Nevertheless, discriminatory policies are not consistent with global economic and technological realities and may be counterproductive in the long run. In the committee's judgment, such policies would be detrimental to U.S. national interests. Given the extent of U.S. global technological interdependence, and the many contributions of the U.S. subsidiaries of foreign firms to the U.S. economy and technical enterprise, it is particularly important that the U.S. market remain open to foreign direct investment and that, as far as possible, such open-market policies be reciprocal.

The committee recognizes that there are many troubling issues raised by the recent growth in foreign control over U.S. industrial assets and the extent to which foreign multinationals draw upon the U.S. research enterprise. It does suggest, however, that it is time for a more multilateral approach to foreign direct investment—an approach that acknowledges the pervasive character and positive contributions of foreign direct investment while seeking to define mutually beneficial "rules of the game" for both transnational corporations and their home and host countries. Good corporate citizenship is becoming ever harder to define as the operations of U.S. and foreign-owned firms become increasingly transnational. Such an effort would do much to reduce tendencies toward technology-oriented protectionalism as well as expand international technology exchange. **Therefore, the committee urges the United States to seek more aggressively to forge multilateral consensus on the mutual obligations of multinational corporations and their home and host governments.**

Greater Uniformity in Antitrust Policy at the International Level

There is mounting pressure on public policymakers throughout the industrialized world to revise or reinterpret national antitrust law or competition policy to fit the realities of global competition and avoid disadvantaging their indigenous firms in the global marketplace. Nevertheless, in the context of the current surge of world foreign direct investment and the proliferation of transnational corporate alliances and mergers, often in already highly concentrated industries unilateral approaches to antitrust regulation and enforcement pose two major hazards.

On the one hand, the relaxation of antitrust requirements by the world's leading economies may increase opportunities for monopoly abuse in certain industries and actually impede technical advance. Although there is at present little evidence of anticompetitive behavior in manufacturing and service industries at the international level, the establishment of alliances among former competitors in certain industries and the rising barriers to market entry in several industries as a result of the spiraling cost of technical advance create an environment in which the threat of anticompetitive behavior is increasingly credible. Despite the potential, if not proven, benefits of interfirm cooperation and collaboration, it is essential that we not lose sight of the fact that competition is a major driver for technical advance and structural adjustment.

On the other hand, there is growing evidence that national competition or antitrust laws are becoming significant obstacles to cross-border mergers and acquisitions that do not undermine competition. Such policy-induced impediments to international competition in the name of enhancing competition may also be harmful to technical advance and economic growth.

Both the danger of anticompetitive abuse by global companies and the costs of protectionist antitrust regulation underline a growing need for greater international cooperation in antitrust policy. Discussions currently under way within the multilateral forums of the GATT and the Organization for Economic Cooperation and Development on this issue warrant greater attention and resolve from all industrialized nations, including the United States. **In summary, U.S. policymakers should strive for greater uniformity in antitrust policy at the international level.**

NOTES

1. Research by Munnel (1990) suggests that public infrastructure is as important as the knowledge stock in stimulating productivity growth.
2. Consider, for example, the Exon-Florio amendment to the Omnibus Trade and Competitiveness Act of 1988, or the spate of bills currently pending in Congress, including the American Technology Preeminence Act (H.R. 4329), Technology Corporation Act of 1990, and others that seek to spell out in legislation specific "special" requirements for foreign-owned or foreign-controlled firms' participation in publicly funded research and development initiatives.

REFERENCES

Council on Competitiveness. 1988. Picking Up the Pace: The Commercial Challenge to American Innovation. Washington, D.C.: Council on Competitiveness.

Munnel, Alicia H. 1990. Why has productivity growth declined? Productivity and public investment. New England Economic Review. (Jan-Feb):3-22.

National Academy of Engineering. 1988a. Focus on the Future: A National Action Plan for Career-Long Education for Engineers. Washington, D.C.: National Academy Press.

National Academy of Engineering. 1988b. The Technological Dimensions of International Competitiveness. Committee on Technology Issues That Impact International Competitiveness. Washington, D.C.

National Academy of Engineering. 1990. Assessment of the National Science Foundation's Engineering Research Centers Program. Washington, D.C.: National Academy Press.
National Governors' Association. 1988. State-Supported SBIR [Small-Business-Innovation-Research] Programs and Related State Technology Programs. Marianne K. Clarke, Center for Policy Research and Analysis. Washington, D.C.: National Governors' Association.
National Research Council. 1990. Ohio's Thomas Edison Centers: A 1990 Review. Commission on Engineering and Technical Systems. Washington, D.C.: National Academy Press.
Pennsylvania Department of Commerce. 1988. Ben Franklin Partnership: Challenge Grant Program for Technological Innovation—Five Year Report. Board of the Ben Franklin Partnership Fund. Harrisburg, Pa.: Pennsylvania Department of Commerce
President's Commission on Industrial Competitiveness. 1985. Global Competition: The New Reality. Washington, D.C.: U.S. Government Printing Office.
U.S. Department of Commerce. 1990. Emerging Technologies: A Survey of Technical and Economic Opportunities. Office of Technology Administration.
U.S. Department of Defense. 1990. Critical Technologies Plan. Prepared for the Committees on Armed Services, United States Congress. March 15.

APPENDIXES

APPENDIX

A

Industry Technology Profiles

I

Aircraft Engine Industry

BRIAN H. ROWE

The current worldwide aircraft engine industry is dominated by three companies: GE Aircraft Engines and Pratt & Whitney in the United States, and Rolls Royce in the United Kingdom. Each of these companies is capable of producing a full line of state-of-the-art engines ranging from small (less than 1,000 horsepower) turboprops/turboshafts to high-performance afterburning military fighter engines to large (more than 20,000 pounds of thrust) high-bypass turbofans. It is in this last category, the high-bypass turbofan used on large commercial transport aircraft, that most of the activity related to the so-called globalization of technology has taken place.

Between the three full-line suppliers and the vast network of subcontractors and component vendors there exists a layer of second-tier players (Table A-1). These consist of several U.S. and foreign companies who have limited whole-engine capability, that is, who are capable of designing, developing, manufacturing, selling, and supporting aircraft gas turbine engines, or major portions thereof, in some but not all segments of the market.

The industry structure is heavily influenced by an extremely long product life cycle. The initial version of a new engine takes four to five years to develop from a well-established technology base, and an engine program, once development has begun, may span more than 30 years before the last engines produced are taken out of service. During this period, the manufacturer usually introduces several major improvements to the engine model family, secures additional applications for derivative versions of the original design, and enjoys a revenue stream from replacement parts that may equal the sales volume of the original engines.

TABLE A-1 Aircraft Gas Turbine Engine Industry Participants

	UNITED STATES	EUROPE	JAPAN
Prime Manufacturers	GE	Rolls-Royce	
	Pratt & Whitney		
Second-Tier Players	Garrett	SNECMA	IHI
	Allison	MTU	KHI
	Textron	Volvo	MHI
	Williams	Fiat	
	Teledyne	Turbomeca	

As engine systems become more complex and expensive, the success of an engine program has become increasingly dependent on product support. Once an engine is put in operation, customers expect that the cause of service problems will be quickly identified and that redesigned parts will be readily available. Also, because growth versions of aircraft are usually heavier and have more demanding performance requirements, the engine manufacturer must be capable of improving the original design to produce more thrust without sacrificing interchangeability with earlier models. Together, these growth and reliability requirements dictate that a relatively high level of R&D spending continue well beyond initial certification and throughout virtually the entire production life of the engine.

In the past decade, alliances have been established between the prime manufacturers and the second-tier companies, and among the second-tier companies themselves, to share technology, reduce fixed costs, and increase market access. Typically, one of the prime manufacturers establishes a long-term business relationship with one or more of the second-tier companies to develop a new engine, which is then sold in regions or market segments where the partners enjoy some type of competitive advantage. At a minimum, in return for providing some of the requisite development funding or effort, these second-tier partners are entitled to manufacture some of the major components or subassemblies of the engine, both for new whole engines and for the spare parts, which are replaced throughout the service life of the engine.

The industry's competitive intensity has been widely publicized; it has resulted in lower product cost to the customer, more frequent improvements in product performance and reliability, and shorter intervals between major advances in technology. The alliances formed between the prime manufacturers and the second-tier companies help to reduce the growing financial burden associated with increasing worldwide competition without jeopardizing their technology leadership.

For GE and Pratt & Whitney, the direction and pace at which critical technologies advance is heavily influenced by U.S. government requirements for both applied research and specific military engine development programs. Both companies have engineering functions that spend approximately $1 billion on research and development annually, roughly divided between military (government-funded) and commercial (company-funded) applications.

In addition to being the principal source of technology funds, the U.S. government imposes tight export controls on what are deemed to be the most advanced technologies, not necessarily limited to those contained in the latest military systems. Restrictions imposed by security clearance requirements for personnel working on classified military programs practically exclude using engineers who are foreign nationals. A government policy requiring that dependence on foreign sources for raw materials or finished parts be kept to a minimum is somewhat more flexible.

To remain competitive, each of the U.S. prime manufacturers maintains its own full set of materials and the design and manufacturing process technologies that are needed for developing and producing new engines across the full product spectrum. Except as required by the U.S. government in case of dual production sourcing, there is no sharing or exchange of technology between the two companies, and yet both companies are viewed as being essentially at technical parity, as is Rolls-Royce. Consequently, the strongest competitive advantage accrues from either having the earliest availability or being able to maintain a sole-source position in a successful aircraft program.

Even though finished parts supplied by vendors constitute roughly 40 percent of the typical engine's manufacturing cost, the prime manufacturers perform the total design function on these parts and require their suppliers to adhere to the same stringent manufacturing standards as exist in the prime manufacturers' own factories. However, the industry's sourcing structure for purchased parts does little to isolate one prime manufacturer's process technology from the other's: 24 of GE's 25 largest suppliers also sell similar components to Pratt & Whitney, and several of the second-tier companies have alliances with more than one of the prime manufacturers.

In a major engine program, the role of the second-tier companies lies somewhere between the prime manufacturers and the vendor network of finished parts suppliers. In return for incurring a portion of the development expense, the second-tier partners usually receive the technology needed to use the latest machines, production tooling, and process technology, which enable them to produce complex parts from what are generally unique and difficult-to-work materials. In those cooperative agreements in which the second-tier partners are also responsible for the design of the parts they will produce, there is some transfer of limited aerothermodynamic and structural design technology from the prime manufacturer. However, the prime manufacturer is able to prevent any erosion of his technology leadership by retaining control over the design of those engine components that represent the greatest technical risk, and the integration of all component designs into the total engine system. In addition to market access, the second-tier partner gains current, component-specific technology (mainly in manufacturing processes but increasingly in design), as well as the scale benefits of greater production loading. As these smaller partners gain experience across several different engine programs, limited but valuable technology begins to flow back to the prime manufacturers (see Table A-2).

The key to maintaining technology leadership in the U.S. aircraft engine industry is a stable, synchronous relationship with the U.S. government. A national policy that would seek to preserve leadership by compelling U.S. high-tech companies to deny others access to their technology may be self-defeating. In aircraft engines, U.S. leadership has been built upon a healthy balance of sustained public and private investment in a vigorous research and development function staffed by competent, imaginative people. An environment that supports the activity of an entrepreneurial technologist and rewards risk taking will nurture the continued development of leading-edge technology.

The accumulation of a series of interrelated new or advanced technologies, coupled with the perception of a market opportunity, can trigger the initiation of a new engine development. As the product-specific development team takes on its task of integrating the new concepts into a total propulsion system, a strong, well-funded applied research function moves on to new challenges and concepts, seeking major improvements or even another new system for initiation several years away. The span and complexity of this process create a time buffer that separates the leading-edge technology from that which is being incorporated into engines in near-term development or production. This inherent natural protection is superior to any restrictive public policy, provided the impetus for advances in technology is maintained.

There is a strategic-defensive reason why GE and Pratt & Whitney should continue to share their technology with the Europeans and Japanese. If they become dissatisfied with the existing relationships, they might be driven to form a true non-U.S. alliance—possibly led by Rolls-Royce—which would have both the resources and the market access to pose a serious challenge to U.S. industry leadership, as Airbus Industrie has done in large commercial aircraft.

There is a vague, judgmental distinction between giving away too much technology and yielding too little; either extreme can weaken U.S. industry.

Today's reality is that alliances are vital to being a world-class competitor, and prudent, controlled technology transfer is essential to strong, mutually-beneficial alliances. Neither of these is as threatening to U.S. leadership as would be our failure to support—with funding and people and public policy—and insist on broad, bold initiatives that advance critical aircraft engine technology.

TABLE A-2 Aircraft Engine Technology Profile

Current Technologies	Future New Aircraft	Critical Technologies
Aerothermodynamics design U.S. leads in critical hot section design	High-performance fighters	Very high temperature turbines, combustors Vectoring, ventral nozzles Low observables
Structures design U.S. leads but Europe gaining	High-speed transport	Short supersonic, mixed compression inlets Low-emission combusters
Controls U.S. leads in applications, but Japan taking the lead in hardware		Low-noise exhausts Advanced integrated controls
	Subsonic transport	High pressure, temperature core components
Systems integration U.S. has slight lead on Europe, more on Japan		Low drag/weight nacelles High-efficiency fans High-temperature composites
Materials U.S. leads but Europe & Japan passing U.S. in nonmetallics	All	Advanced manufacturing processes Testing facilities, methods
Manufacturing processes U.S. leads in technology, but Europe and Japan implementing faster		

II

Automotive Industry

W. DALE COMPTON

The automotive industry has been transformed in the past decade. Whereas its design and manufacturing facilities were once located near the markets that it serves, the industry now offers products that are designed and manufactured in a dozen or more countries and are marketed in hundreds of countries. The conversion to a world marketplace has created a competitive environment that rewards product quality, product reliability, low cost of ownership, and reliable service, irrespective of where the product is manufactured.

From an international perspective, the automotive industry is technologically more homogeneous than might be surmised from a casual examination of the performance of various manufacturers in the marketplace. Recent comparative studies of the industry in the United States and Japan strongly suggest that the competitive advantage enjoyed by the Japanese does not arise from a technical advantage. Similarly, the technology used by the European manufacturers does not differ substantially from that used by U.S. manufacturers. Neither would a significant difference be found between the level of the technology used by the automotive industry in Brazil, Korea, Taiwan, Italy, Australia, or Canada and that used in the United States, an observation that is not surprising since U.S. companies are strong participants in many of these markets. This homogeneity does not mean, however, that the industry of a particular country may not be technologically superior in a specific area, for example, Brazil's use of alcohol fuels. Although this superiority tends to be the exception rather than the norm, it is important to recognize that regional differences in the marketplace can also strongly affect the technological level of the products offered in those regions. This can be seen in the emphasis on characteristics such as high-speed performance and fuel economy which are strongly influenced by local customs or government regulations.

One must conclude, therefore, that the current competitive advantage enjoyed by some manufacturers, for example, the Japanese, results not from better technology but from a better management of their overall system. This includes, of course, the way that they use technology, their continuing emphasis on quality, and the continuous improvement of all operations, and in some instances, lower costs. For this committee, the following three key questions seem relevant to the discussion of "engineering as an international enterprise" as it relates to the automotive industry. Why did the homogeneity develop? Is this technological homogeneity likely to change with an accompanying increase in domination of the world industry by companies located in one geographic area? What impact will these trends have on the engineering capability of the United States?

The answer to the first question is a direct consequence of an industry structure that can be roughly described as a combination of (1) large multinational companies with design, manufacturing, and marketing activities in many countries; (2) national companies that design and manufacture products in one country but market these products worldwide; and (3) a variety of business arrangements that involve joint ventures, minority ownerships, and purchase agreements for components and vehicles. In this regard, each of the major U.S. companies owns equity in one or more Japanese companies. With regard to national companies, there are local companies such as Citroen and BMW as well as subsidiaries of multinationals that have existed for decades and are often treated and considered by the host populace as local national companies. As examples of the latter, both Ford and General Motors have subsidiaries in Europe, Asia, Central America, and South America that have design, engineering, and manufacturing capability. One should conclude, therefore, that the globalization of the automotive industry is not a new development. What is new is the capability that the industry now has to use these operations, irrespective of their location, to design and manufacture products for sale to customers who have an option to choose from a variety of products made by companies located in all regions of the world.

The globalization of the industry is probably a necessary but not a sufficient condition for homogenization of the technology. The presence of manufacturers in a wide variety of markets, the capability to acquire and analyze the products of all manufacturers, and the opportunity through various business relationships to share technology suggests that the current homogeneity of capability is a logical consequence of this diversity of location and business arrangement. International professional societies, such as the Society of Automotive Engineers, have been important contributors to the technical homogenization of the industry. Although the manufacturing processes used by the various manufacturers are somewhat less homogeneous than is the product technology, many manufacturers use common vendors for their processes. This contributes to the homogenization of the manufacturing technology. Joint manufacturing arrangements also contribute to homogenization. The consequence of the current industry structure is that it would be difficult, if not impossible, for one company to dominate all others technologically for an extended period of time. It must be recognized, however, that any company can enjoy a short-term *technological* advantage through the introduction or use of a new or unique product or process technology.

If this homogenization of technology reduces the competitive advantage that a company may be able to exercise through technology, where does the competitive advantage lie? Clearly it rests with the company that can most effectively use a broad range of technology to produce a product of outstanding quality, at the lowest cost, with the shortest delay between the concept and the market. With a homogenization of the technological base, a company must excel in its execution rather than depend on a technological advantage. With regard to management practices, the homogenization of the industry does not directly offer a ready means of transferring information about new operating practices and procedures. Even though Ford, Mazda, GM, and Toyota have all learned important lessons from their joint programs, the introduction and dissemination of new concepts throughout a large organization can be very difficult and require much time and effort. Thus, the advantage that some companies have in effectively managing their systems and their technologies remains a principal competitive force.

Is the current structure of the industry destined to change in the next decade? It seems unlikely. Recognizing that the industry probably has an excess of capacity to supply the expected world market, one can expect that contraction will encourage more alliances and joint relationships among the surviving members. Responding to "local content laws" is likely to force manufacturers to distribute their manufacturing facilities geographically, thus precluding the concentration of facilities within the borders of one country. Location of facilities will continue to be made according to the opportunity to reduce costs and expand markets. Although the consequence of this trend is likely to be a continued shifting of facilities from one country to another, the result is not likely to have a significant effect on the homogenization of the industry.

Does the continued movement of facilities from the United States to other countries suggest that the United States will lose its capability to compete in the world automotive market? The answer, of course, depends on the degree to which overseas sources will displace U.S sources. If the industry were to cease to design and manufacture vehicles in the United

States—an admittedly extreme situation—the United States would lose the infrastructure, including the supply base, necessary for a viable industry. Once lost, it is likely that regaining it would be impracticable. At the other extreme, it would be unwise to suggest that the U.S. auto industry should not take advantage of the many opportunities that exist for developing joint business arrangements with foreign companies. Such arrangements often afford the U.S. industry access to foreign markets, provide a basis for sharing the burden of investment, and provide a means by which technology can be assessed and evaluated.

Should a decision by a United States manufacturer to locate a new engine or transmission plant overseas be cause for alarm? If it were one of many such plants that exist in the United States, the chances are slim that this decision would lead to a serious decline in the technical capability of the United States-based industry. If it is one of a few, the answer could be different. Because of the dynamic and complex nature of the system, one cannot easily establish a priori a standard that indicates that a fixed level of capability is essential. The best one can do is to examine continually the many issues that determine the viability of an industry and to assess trends as they occur. Recognition of this fact led the NAE Committee on Technology Issues That Impact International Competitiveness (1988) to the following recommendation:

> Before joint government-industry actions are undertaken, an important early step must be sound analyses of all aspects of the problem, including an understanding of the technological status of critical sectors of U.S. industry, the implications of emerging technologies for the health of engineering and technology in all sectors of U.S. industry, and deficiencies in the technological infrastructure of particular sectors. . . . A small activity, perhaps located outside the structure of the government, staffed by highly qualified analysts who are keenly aware of industrial problems in detail, could be of great value.
>
> With analyses of the type described above, the government would be better prepared to respond to industry initiatives.

A few general observations regarding the automotive industry also are pertinent to the role that the industry infrastructure plays in the development and use of technology.

First, the Japanese industry has long developed a closer working relationship with its supplier base than has the U.S. industry. This has created a feeling of belonging to the "family" that has contributed greatly to the capability of the Japanese industry to implement just-in-time systems, improve quality, and introduce new technology in components and subsystems. Although U.S. manufacturers are making progress in achieving some of the same relationships with suppliers, the Japanese industry continues to benefit greatly from a long-standing tradition in such relationships. One should note that this represents a form of vertical integration without the actual legal or direct financial commitment that would be required of "true" integration.

Second, the linkages among the various engineering functions in the value-added chain of an automotive company are extensive. One need only consider the continuing advantage derived by Japanese companies through the shorter time that they require to bring a new vehicle design to the marketplace to recognize the importance of this linkage. In this example, the issue is whether the Japanese companies possess this advantage through technology or whether it is a result of an improved management system. I believe that the evidence strongly suggests that it is the management system that gives them the advantage, such as early consensus building, few changes in objectives, and few engineering change orders.

Third, there are certain capabilities that every manufacturer must have if it is to be competitive. For example, system integration remains a key element, whether a manufacturer is vertically integrated or depends on a wide supplier base for components and subsystems. Successful system integration depends on a broad expertise in essentially all technologies of the vehicle. One cannot expect to be competitive by buying major components and subsystems as "black boxes." In this sense, a broad capability in a wide range of technologies is essential.

Finally, one should note that as competition becomes more intense, many of the factors that have been traditionally used to provide product differentiation will cease to function in this way. The ability to provide a recognizable value to the customer through the application of technology may ultimately be a key determinant of success in the marketplace. Consequently, the successful use of technology will likely become a critical determinant of competitive success in the future of the automotive industry.

REFERENCE

National Academy of Engineering. 1988. The Technological Dimensions of International Competitiveness. Committee on Technology Issues That Impact International Competitiveness. Washington, D.C.

III

Biotechnology

ELMER L. GADEN, JR.

The scope and character of most of the industrial areas considered by the NAE study committee may be defined in terms of the products furnished or the services rendered. Indeed, many are formally defined under the Office of Management and Budget's "Standard Industrial Classification" system.

"Biotechnology" presents a significantly different picture. It is not identified with a specific group of products or services. Rather it encompasses a broad range of activities and processes in which living cells—or materials produced by them—are used in a technological mode. It therefore comprises a set of *techniques* rather than products and services. In fact, many knowledgeable persons strongly object to the use of the term "technology" for what is, in many cases, primarily a set of laboratory techniques, powerful and sophisticated as they may be. Note the following paragraph:

> Biotechnology is not an industry *per se*, but rather an array of technologies that can be applied to a number of industries. These technologies include: molecular and cellular manipulation, enzymology, X-ray crystallography, computer modeling, biomolecular instrumentation, industrial microbiology, fermentation, cell culturing, and separation and purification technologies. (U.S. Industrial Outlook, 1989)

In this sense "biotechnology" serves a wide spectrum of industries and commercial activities:

- Agriculture and animal husbandry
- Food production and processing
- Health care
- Chemical production; both commodity and specialty products

- Textile manufacture
- Mining and mineral processing
- Waste treatment and disposal; resource recovery

DEFINING "BIOTECHNOLOGY"

In the most comprehensive terms, biotechnology may be thought of as comprising all aspects of the technological exploitation and control of living systems. Such a broad interpretation has the virtue of incorporating a wide spectrum of familiar activities of great economic importance; agriculture, animal production, food preservation, brewing, and production of natural rubber and paper are examples. Furthermore, biotechnology—in these terms—is hardly new. Indeed it is one of the oldest of mankind's technological activities.

It is unlikely, however, that an appreciation of these traditional practices is a sufficient basis for including "biotechnology" within the committee's purview. Rather, it is the "new" biotechnology, based primarily on the deliberate manipulation of genetic material, that is of most interest here.

Nevertheless, despite the promise of the "new" biotechnology, it will not serve the purposes of this study to ignore the traditional aspects of this field. To that end let us return to the general definition offered above—"biotechnology comprises all aspects of the technological exploitation and control of living systems"—and briefly identify the various activities it embraces as follows:

Biotechnology

- Control
- Exploitation
  - Extraction
  - Bioprocesses

Control: Technologies that either (1) restrict or control the activities of a wide range of organisms in an equally wide range of environments or (2) eliminate all life forms ("sterilization") are control technologies. They include the following examples:

- In agriculture—fungicides, herbicides, insecticides
- Food preservation
- Water purification
- Preservation of biologically labile materials:
  Natural—wood, cotton, etc.
  Synthetic—hydrocarbon fuels, polymers, etc.

Extraction: Nature has provided an infinite variety of useful molecules and molecular composites; we need only extract and purify them. The methods employed vary from simple physical separations with little or no effect on the molecular species encountered to more traumatic treatments involving significant chemical transformation. Some examples are as follows:

- Extraction of tannins, dyes, medicinals, and oils and fats from plant or animal tissues
- Sugar (sucrose) from cane or beets
- Latex (rubber) from trees or shrubs
- Starch from corn, wheat, etc.
- Cellulose fibers (for papermaking) from wood or grasses
- Charcoal, liquid and gaseous products by pyrolysis of wood

Bioprocesses: Complete living systems (cells or tissues) or their components (enzymes, membranes, chloroplasts) are employed in a directed and controlled manner to bring about desired physical or chemical changes. Examples include the following:

- Production of cell matter—mushrooms, baker's yeast, etc.
- Production of cellular components—enzymes, nucleic acids, etc.
- Chemical products ranging from "specification" (e.g., ethanol) to "performance" (biopolymers) types
- Pharmaceuticals
- Waste disposal and conversion
- Extraction of minerals from ores and collection of metal ions from dilute solution

THE "NEW" BIOTECHNOLOGY

The "new" biotechnology rests upon (1) rapid expansion of our understanding of the mechanisms by which genetic information is stored, transferred, and transformed and (2) the development of laboratory methods for the deliberate manipulation of genetic material. Furthermore, it is truly "new." Specific proposals for commercial exploitation of these advances in basic science were first made about 15 years ago. The first significant product—human insulin produced by a bacterium developed by recombinant DNA methods—has been marketed for about 8 years.

Several of the most visible applications of the "new" biotechnology have been in the pharmaceutical area—insulin, human growth hormone, tissue plasminogen activator, etc. As a result, discussions of the potential for biotechnology often convey the impression that it is a major contributor to the large number of new pharmaceutical products (drugs) introduced each year. In fact the overwhelming majority of these new drugs are produced by chemical synthesis, not by biological methods.

ONGOING TECHNOLOGIES AND COMPETENCIES

Key Elements of Bioprocess Technology

If we restrict ourselves to the "new biotechnology" and, even more specifically, to its bioprocess component as defined above, we find that the key elements are as follows:

1. Biocatalyst. The "biocatalyst" is the agent—microorganism, cell, enzyme, etc.—responsible for catalyzing the desired chemical change or synthesizing the product of interest, such as an antibiotic or a biologically active protein.
2. Substrates. The ingredients or raw materials on which the biocatalyst acts or a microorganism is grown are called substrates. These ingredients must include energy (carbon) sources (typically a sugar such as glucose), nitrogen sources, phosphorus, growth factors, and a variety of micronutrients (trace components), small in amount but vital in function.
3. Conversion system. Conversion systems include the process equipment and controls in which the biocatalyst and substrate interact under conditions contrived to provide the "direction and control" referred to in the definition of bioprocesses given earlier.
4. Separation and purification system. Separation and purification are provided by the process equipment and controls that permit recovery of the desired product or products from, first, the large amounts of water ordinarily present and, second, undesirable by-products and contaminants.

Biocatalysts

Of these four primary technological elements, the first—the biocatalyst—is clearly the key; without it nothing is possible. These agents have therefore been closely guarded and, in consequence, have been the focus of a few bizarre cases of industrial theft. The 1980 Supreme Court ruling (Diamond v. Chakrabarty) that microorganisms may be patented under existing law cast a new light on this matter. Nevertheless, the obvious difficulties in detecting and policing infringement ensure that secrecy continues to dominate industrial practice.

In this context a recent regulatory initiative by the U.S. Patent Office should be noted. Current rules require the deposit of "essential biological material"—that is, the microorganism itself—in a suitable depository. The

Culture Collection of the Northern Regional Research Center, U.S. Department of Agriculture, in Peoria, Illinois, is commonly used for this purpose. The Patent Office's new proposal would permit an applicant, under certain circumstances, to provide facts and data interpretation to support an application without depositing the organism. The Patent Office is also considering some post-grant restrictions on public access to deposited organisms. Current practice allows anyone to obtain the deposited organism from the depository once the patent is granted.

Also fundamental to the questions of technological interdependence and competitiveness is the manner in which the "new" biotechnology was developed. Most of the novel techniques on which it is based, for example, recombinant DNA, protoplast fusion, and hybridomas, were developed in the United States. Hybridomas, which were first developed and patented by a British national, Milstein, were a notable exception. It was at first assumed that this situation would guarantee U.S. dominance in the field. This has not been the case and the reasons are clear.

First, virtually all of the basic research underlying the "new" biotechnology was carried out in universities and medical research institutes and was supported by federal government funds. Consequently, much of the basic work was both in the public domain and freely and widely published. There were—and are—few proprietary positions.

Furthermore, once worked out, the laboratory methodologies required are relatively simple, albeit tedious. As a result, even developing countries with relatively young science establishments are active.

Separation and Purification

The second critical technological element is the separation and purification system. For many years the attention of the research community, especially its academic component, was focused almost exclusively on the conversion phase. Great improvements were realized, but over the last decade—and probably longer—they have tended to be peripheral and minimal in economic impact. Only for about 10 years has attention moved to separation and purification. Many believe that it is in this arena that key battles will be fought for future economic advantage.

In this connection, a recent development is worth noting. One of the major U.S. pharmaceutical firms, in its first attempt to penetrate the Japanese market, has found that Japanese quality standards are significantly higher than those in the United States. Purity and "elegance factor" (speed of dissolution, clarity of resulting solution, etc.) criteria imposed by the Japanese pharmacopoeia apparently exceed those of the U.S. pharmacopoeia by significant margins.

THE "BIOTECHNOLOGY" INDUSTRY

At its inception the "new" biotechnology epitomized the "high technology" concept. Its appearance on the national and international scenes coincided with rising concern over the decline of traditional manufacturing industries. As a result it rapidly captured the attention of the media, the general public, and government leaders as well as venture capitalists eager to be in on the ground floor of a "new industrial revolution." It should be noted that virtually every Western European country as well as Canada, Japan, and several international agencies, such as the Organization for Economic Cooperation and Development, established special commissions to consider and recommend policies and plans for exploiting the "new" biotechnology. There was no comparable effort in the United States until the Office of Technology Assessment of the Congress published its *Commercial Biotechnology: An International Analysis* in 1984.

One special characteristic of this era was the appearance, primarily in the United States, of a plethora of small "biotechnology" or "genetic engineering" companies. Most of these were built around persons from universities or research institutes who were active in the scientific work underlying these developments. A few have been quite successful, but many have fallen by the wayside or been acquired by established pharmaceutical, food, or chemical companies. At first these larger companies lacked the specialized scientific base for this new field, but they have rapidly acquired it. Furthermore, they were familiar with the development of high-technology products, possessed the marketing networks needed to sell them, and knew their way around the complex regulatory procedures governing the health-care and agricultural businesses. All evidence indicates that this pattern is likely to continue.

A major difficulty in the development of the new biotechnology companies is that investors have been led to expect the spectacular. Instead, growth has been slow and steady. Furthermore, the commercial performance of "biodrugs" has been disappointing because they have been found to be far less useful than originally hoped. Even Genentech, the most successful of the "genetic engineering" firms, is facing difficult times because of the failure of its t-PA (tissue-plasminogen activator) to demonstrate significant advantages over competing drugs.

EMERGING TECHNOLOGIES

Here we must return to the original argument, namely, that "biotechnology" does not constitute an industry in the usual sense. Rather it comprises a group of techniques, processes, and procedures applicable in various ways to many industries. Underlying these techniques and procedures is a large and rapidly expanding base of scientific knowledge, virtually all of it freely published and readily available to technically competent individuals and organizations.

The United States clearly continues to be a world leader in developing these techniques and procedures, and the leader in developing the science base on which they rest. Large and effective research efforts are in place in both the public and the private sectors, and new developments are reported almost daily. The difficulties that we face in these areas are therefore not primarily technological.

Some claim that excessive public regulation is an impediment. Given the immense public support that biotechnology has enjoyed and the errors, happily few but no less tragic, that have been encountered, it is difficult to substantiate this claim. A primary problem is that technical advances and many of the economic opportunities they offer are very short-lived. Furthermore, foreign countries may impose regulatory barriers that slow down market penetration sufficiently to permit local competitors to move in when acceptance has been achieved—often at the expense of the U.S. innovator!

In conclusion, this is an area in which the United States has been successful in both innovation and execution. Despite this, biotechnology's contribution to our international position has not been impressive. Clearly the problems lie elsewhere.

BIOTECHNOLOGY AND "ENGINEERING AS AN INTERNATIONAL ENTERPRISE"

The emphasis throughout this profile has been the close relationship between the practice of biotechnology—the "biotechnology industry"—and the science base on which it rests. Implicit in this point is the lack of significant engineering contributions. Despite claims (primarily by university engineering faculty in search of grants!) that future progress could be limited by a lack of "engineering skills," there is little credible evidence to support this contention. Biotechnology is and will likely continue to be "science-limited" rather than "engineering-limited" for the next decade at least.

IV

Chemical Process Industry

EDWARD A. MASON

The chemical process industry has developed over the past 90 years and is now one of only two U.S. industries with a positive net balance of payments for exports and imports. Much of its present structure grew out of advances in synthetic chemistry in Germany late in the last century. German chemists conceived new chemical syntheses, but the scale-up to commercial operation was carried out by mechanical engineers who were not well grounded in chemical principles. Thus, many early German plants were merely scale-ups of the batch laboratory syntheses designed by the chemists.

It was American engineers, many of whom studied chemistry in Germany in the early decades of this century, who developed the basis for chemical engineering education and practice. These chemical engineering pioneers developed the concepts of unit operations, such as filtration, distillation, heat transfer, fluid flow, and crystallization, which are applied in the commercial production of a wide variety of chemicals in continuous processes.

Consequently, the United States became the leader in the education of chemical engineers from around the world and in the engineering of chemical reaction processes. A wide variety of engineering skills and disciplines are now employed in the development, design, construction, and operation of economically optimized processes for the production of both commodity (large volume, low unit value) and specialty (lower volume, higher unit value) chemicals. From the 1930s to the 1960s, American engineering know-how played a dominant role in the worldwide chemical process industry. Although it is still a major influence worldwide, the growth in demand for chemicals of all kinds around the globe has led to expansion in international chemicals production and engineering expertise. However, American universities are still preeminent in chemical engineering education and research.

All developed countries and many developing countries have several large and small chemical companies producing a wide variety of products. A large fraction of chemical products are derived from petroleum and natural gas. Thus, much of the industry draws on a worldwide supply of feedstocks. In addition to the many independent chemical companies, most petroleum producing and refining companies have large petrochemical operations.

The role of engineering in the chemical process industry is manifold, from the development of a new chemical process plant to its construction and operation. Development of the plant begins with chemical research and development, which define the broad parameters of a new chemical process. Usually this involves research chemical engineers, along with chemists to conceive and develop the chemistry itself. These broad parameters are transferred to a process engineering group, where a conceptual design of the process and plant is developed. Then a central process engineering group does the overall design, including the piping and instrument diagrams. During this process, energy balances are developed, and the dynamics of flow in the system are defined, including requirements for fluid transfer pumps, heat exchangers, reactors, heat recovery, concentration or extraction, and purification. After completion of the piping and instrument design, the work becomes more detailed. Electrical, mechanical, and industrial engineers are brought in where additional considerations of materials of construction, process computer requirements, and process optimization and modeling are developed. Preliminary cost estimates and process optimization are important aspects of the work.

At this point, the scope of the process is defined somewhat conceptually to give the general parameters and sizes of the major equipment in the plant. The standards of the corporation that will own and operate the plant are introduced, and an overall process scope is defined.

Because very few chemical process companies have their own engineering construction design teams, the project scope is sent to contract engineering firms for bids. These firms come back with proposals to build the plant on either a fixed-fee or a cost-plus basis. When the proposals are evaluated, the owner's cost estimates, which have been made internally, are updated and estimates of the economics of profitability are then better defined. The contract engineering firm develops a detailed design in which draftsmen, electrical engineers, civil engineers, mechanical engineers, experts in soil mechanics, and environmental engineers are involved in defining the process and its waste streams and in minimizing environmental impacts.

Once the contract has been approved and management has decided to build the plant, construction engineering follows and a project team is established. The project team members are brought in early to work with the design teams. In addition, plant operations and maintenance personnel are also brought in to ensure that their viewpoints are incorporated into the design and construction of the plant. During the final stages of construction, personnel who will be responsible for operating the plant are brought in to check the operation of individual process units, do pressure testing, and generally confirm the integrity of the system. Maintenance personnel check out instruments, and a training program for operators is initiated.

The entire plant design, as well as the construction operation, are subject to hazard and operability review to ensure operational safety. Fault-tree analysis is used extensively in the chemical process industry.

Once a plant is in operation, engineering continues to optimize its performance and to eliminate bottlenecks so that plant capacity can be increased. In these plant changes, chemical engineers, mechanical engineers, and process control engineers work together. The mechanical engineers are heavily involved in many chemical process plants with rotating equipment, while the electrical engineers are concerned with controls, motors, drivers, and the like.

Computer-aided design techniques are now being used both for electrical and piping layouts and for process design evaluation and optimization. Statistical process control is widely used in quality improvement programs.

An extremely important part of the engineering that goes into the development and construction of a new chemical process plant is that of project management. This function is extremely important to ensure the efficiency and the integration of the project, which involves the efforts of many people from various disciplines.

In the United States, it is generally customary to use U.S. engineering firms and engineering contractors. In other countries, however, the use of American engineers and firms has declined for a number of reasons. Foreign nationals are used increasingly because they know the culture and the local ways of doing business. In most cases, they are good engineers, and in many countries they must be employed. The American owner of the process technology in another country often is not allowed to own a majority of the company and is thereby denied control of the project once it is in production.

Formerly, many American engineers worked overseas. Changes in the tax laws, however, removed what were perceived as advantages for U.S. citizens working abroad, and the cost to American companies increased. This is one reason why many companies that were already experiencing high costs to employ Americans overseas began to pull them back. Thus, it became even more necessary to rely on the indigenous work force for engineers and constructors. Foreign nationals naturally tend to specify and order their own country's equipment, whereas American engineers would be likely to specify and order American equipment for overseas construction.

The American licensor of the process would like to have a technical service agreement, but this is not always possible. Such an agreement would provide for nondestructive testing, corrosion services, and vibration analysis, all of which are needed during plant start-up and operation.

Nowadays, it is rare to have in-house construction management. Engineers who work for the engineering design firms and engineering constructors tend to be somewhat migratory, following jobs from one company to another. Especially critical are good construction managers.

Years ago, there was much discussion about the brain drain from overseas to the United States. Now, as there seems to be a growing shortage of American-born scientists and engineers, it is likely that the chemical process companies, as they begin to expand, will go overseas again to recruit. The need to have an overseas presence, including foreign nationals on overseas ventures, is becoming increasingly critical and will become more so in the 1990s as the European Community consolidates.

In summary, the chemical process industry depends on skills in thermodynamics, fluid mechanics, computational computer science, modeling, materials of construction, chemistry, physics, automatic control, electrical engineering, civil engineering, environmental engineering, safety analysis, procurement, hazardous waste management, vibrational analysis, quality control and assurance, project management (people skills and ability to integrate a variety of functions). Because of the increasingly international character of the chemical process industry, the United States no longer holds a unique position in engineering for the industry.

V

Computer Printer Industry

DONALD L. HAMMOND AND WILLIAM J. SPENCER

The modern electronic computer represents a several hundred billion dollar global business. Even more important, computing capability has inundated almost every aspect of work, play, and education. Today's computers depend on a variety of technologies including integrated circuits, software, communications, and, of growing importance, the ability to access the large amounts of information processed by computers. A key part of the human interaction with computers is through printers that provide textual or graphical information from the entire range of computers from desktops to large mainframe and supercomputers.

The major change in computing today is a move from central processors to distributed processing usually in the form of a desktop personal computer or workstation often networked both locally and globally. This move towards decentralized computing is also reflected in the increased presence of personal printers. Today, desktop printers represent a multibillion dollar business that is continuing to grow at a rate of about 10 percent per year.

Two technologies are beginning to emerge as dominant in the desktop printer business. The first is laser printers. These quiet, plain-paper devices are built around electrophotographic concepts that were pioneered by Xerox in the development of plainpaper copiers. The light-lens exposure system in a copier is replaced with a solid state scanning system to produce high-quality text on plain paper. The second emerging technology is for lower cost, lower speed printers built around ink jet printing technology. The simultaneous and independent invention of thermally driven ink jet printers by Canon and Hewlett Packard enables high-quality, very low cost printers that provide letter quality print on plain paper in either black and white or multicolor.

There are, of course, a variety of other print technologies including impact, thermal transfer, other types of ink printing, and a variety of exposure techniques other than laser for electrophotographic printers. Impact printers, both daisy wheel and dot matrix, have dominated the print business in the past for both desktop and centralized printing. However, these markets are rapidly changing and it appears clear that ink jet and laser printers will be the key to computer hardcopy output in the future.

Currently, laser printers are made in the United States, Japan, and Europe. The manufacturers of laser printers are usually companies with strong technological capability in electrophotography such as Kodak, Canon, Ricoh, Siemens, and Xerox. The high-speed laser printing business is dominated by European and U.S. suppliers. The low-speed, ten pages per minute and less, are dominated by Japanese suppliers. The total market volume both in bands above ten pages per minute and below ten pages per minute are in the thirty billion dollar range.

The newer thermal ink jet printing area is currently dominated by two suppliers: Hewlett Packard and Canon. This position is bolstered by strong patents in ink jet technology, effective distribution channels, and early market presence. Technologies supporting these two printer areas are divided into current capabilities and emerging capabilities. In the current capabilities, the following technologies are considered significant.

Current Technologies

Solid-State Lasers—Japanese Leadership

Early laser printers used gas lasers to expose photoreceptors in a raster scanned mode similar to black-and-white television. Gas lasers still dominate high-speed laser printers; however, for the smaller, lower speed printers (about ten pages per minute or less), solid state laser diodes have dominated. Solid-state lasers provide higher reliability, lower cost, and provide the flexibility for building small printers. The use of solid-state laser diodes required the development of new photoreceptors that were sensitive to the infrared frequencies usually available in solid-state lasers. Solid-state lasers, as well as other optoelectronic technologies, are currently dominated by the Japanese. Except for a few special applications such as high power, all laser diodes and low cost raster scanning systems come from Japan.

Ink Jet Technology—U.S. Leadership

The ability to eject ink droplets at kilocycle rates from thermally driven heads has opened a new type of printing capability. Initial discoveries were made almost simultaneously in Japan and the United States. Major volume leadership resides in the U.S. because of design emphasis on very low cost, disposable print heads which contain all of the critical ink jet technology. Major investments in understanding the technology, in designing for manufacturability and in automating for high volume production (currently 700,000/month) has resulted in highly reliable, low selling price ($10 to $15), disposable heads that print with 300 pixels per inch. While color printers using thermal ink jet technology are newer, they benefit from these same advantages and are finding similar market acceptance. That leadership is transitory and will require continued investment to maintain U.S. leadership.

Manufacturing Costs—Japanese Leadership

The Japanese have assumed leadership in the manufacturing of most consumer products. Leadership depends on a tightly networked system of materials and components suppliers, and just-in-time delivery techniques. Manufacturing leadership is a result of heavy investment over a long period of time in Japan. The U.S. is catching up in some areas, but overall, the Japanese maintain a significant leadership in manufacturing costs, quality, and the ability to quickly turn around new manufacturing designs.

Design Capability—Japanese Leadership

Closely coupled with manufacturing leadership is the ability to rapidly design low cost, high-quality, highly reliable products. Again, Japan has the leadership in this area as a result of heavy investment over a long period of time. There are, however, exceptions such as the thermal ink jet example. In this field as in others, Japan benefits because design and manufacturing are generally considered more prestigious engineering activities than in other parts of the world. Design leadership is a major competitive advantage.

Printing Materials—U.S. Leadership

In the understanding and manufacture of inks, ink-paper interactions, photoreceptors, toners, development systems, and other ink and electrophotographic materials, the U.S. holds a leadership position. This position is eroding in areas such as low-cost organic photoreceptors and single component development systems. The United States must regain leadership in these key materials areas if it is to continue to play a leadership role in printing technologies, especially for desktop printers.

Software Applications—U.S. Leadership

The development of printing applications, page description languages, and other systems applications have been an area of U.S. leadership. Again, this leadership is under heavy competitive pressure from large software investments in Singapore, India, Japan, and other emerging countries. The Sigma Project in Japan offers a particular challenge to U.S. leadership. The purpose of this project is to develop and prove methods for software engineering and the ability to share software programs between universities, industry, and government laboratories throughout Japan.

EMERGING TECHNOLOGIES

Color—U.S. Leadership

As PCs and workstations become more sophisticated, color has become an important capability. Color printers are available in a number of technologies. Color laser printers are currently on the market only from Japan. Color ink jet printers are becoming available from a number of sources. While the Japanese have made early market entry in most areas of color printing, the major advantage of the disposable color ink jet cartridge has given a significant cost (selling price about $1,400) and performance advantage to printers from the United States and has resulted in a volume of 8,000 per month or many times the sales of color ink jet printers from Japan during this early period in the emergence of the high volume color printing market. Just as color television grew to dominance in the consumer television market, it is clear that color printers will also play a dominant role in the printing market. It is essential that the United States retain and strengthen its capability in these color printing technologies if it is to retain a significant position in printing.

High-Speed Printers—U.S. Leadership

High-speed printing is usually done by laser printers built around high-speed plain paper copiers. Leaders in this area are Kodak, Siemens, and Xerox. Again this leadership position is eroding due to the movement of Japanese printers from low-speed to moderate and higher speed machines. The key technologies here are the ability to handle paper at high speed, printing materials, and the economical design of these complex machines.

Scanning—Japanese Leadership

In addition to printing information from computers, it is important to be able to input information into a computer. Scanners are becoming more widely used as a way to move from paper-based information into the electronic information realm. The hardware for scanning comes principally from Japan in the form of either charge-coupled devices or amorphous-silicon scanners. The United States has an edge in optical character recognition. Future development of scanners will include the ability to go to vectorized graphical images, retain font information in scanned text, and, generally, the ability to handle complex scanned documents.

Facsimile—Japanese Leadership

A personal workstation with a scanner and printer provides an important communication device. The scanner and printer provide a facsimile capability. Optical character recognition permits the movement between the FAX world and the electronic mail world. The workstation is the key interface to the filing and retrieval of information. The combination of workstation, scanner, and printer gives the ability to do communication, filing, retrieval, and processing, all from a desktop facility. It's not difficult to envision the merger of workstations and facsimiles to provide an information appliance for individuals that could be networked as an important tool for information sharing and joint projects.

TECHNOLOGY SUMMARY

There are some important conclusions that can be drawn from these brief looks at the technology in the computer printing business. These include:

1. There has been a movement of U.S. technology to Europe and Japan, particularly in the field of laser printers.
2. The loss of market share in the printer business has led to a loss of design and manufacturing jobs to Europe and Japan.
3. These technologies represent large current markets and markets which are continuing to grow rapidly.
4. The future merging of input/output devices with workstations represents a major threat to U.S. manufacturers of workstations and personal computers.

VI

Construction Industry

MILTON LEVENSON

The U.S. engineering/construction industry is basically an assembly industry. While it does major amounts of engineering, much of it innovative and original, it is almost all in the nature of using conventional basic technology in the application of materials, equipment, or components developed by others. In this century, the U.S. construction industry's large-scale penetration of world markets has been based more on its ability to manage large, complex projects and control quality than its leadership in specific technologies or technical disciplines. The fact that the United States was among the first to develop a comprehensive system of construction codes and standards also helped U.S. companies expand overseas, since many countries lacking their own codes and standards adopted those of the United States rather than develop their own. In short, technology has not traditionally figured as a major arena for competition within the international construction industry. However, trends of the last few decades suggest that the technological dimension of global competition in the industry is becoming increasingly important, particularly as non-U.S. firms expand their presence in the United States and other foreign markets.

The primary source of technological innovation in the construction of new facilities has been the manufacturers and vendors of construction material and equipment. Since the construction industry has always done worldwide procurement for major projects, its technology base has long been international in character. In recent decades, non-U.S. vendors and manufacturers of material and equipment have increased significantly their presence in world markets, often at the expense of their U.S.-based competitors.

In the process, foreign firms' influence over the industry's technological development has grown disproportionately.

Today, a major fraction of the new developments in construction technology are of foreign origin. One example is tunneling equipment and technology. U.S. companies are almost totally dependent on foreign companies for this aspect of construction. At present, U.S. companies appear to be leading in the *development* of three-dimensional computer-aided-design-and-drafting (CADD). However, some non-U.S. companies appear to be more advanced in the application of CADD, and unless things change, they will certainly pass U.S. firms in the race to expand CADD to a complete computer-aided-engineering and computer-aided-construction (CAE-CAC) system. Unless major changes occur, U.S. dependence on foreign sources of technology is likely to increase, particularly as capital intensive means such as robots and automation displace craft labor.

The U.S. lag in construction technology is largely a result of three factors: (1) a higher degree of vertical integration in construction industries abroad than in the United States; (2) more favorable contracting practices overseas than in the United States; and (3) a more favorable regulatory environment in many foreign construction markets than in the United States.

International Differences in Industry Structure and its Consequences

For historical and legal reasons, the scope of activities of U.S. construction firms has been largely limited to the integration of design engineering, procurement, and construction functions. In contrast, many foreign construction companies also include manufacturing as part of their overall chain of value-added activities.

A nonobvious effect of differences in the extent of vertical integration between U.S. and non-U.S. construction companies—one that is amplified by the cyclical nature of the industry—is the difference in career paths and commitments of U.S. and foreign engineering staffs. In the United States, many engineers in the construction and design engineering industries are salaried "nomads" moving from company to company as the work moves with minimum commitment to follow-through or improvement. By contrast, in several European countries and in Japan engineers are involved in much of the lifetime of the facilities they design and build.

Linkages between engineering functions in the U.S. construction industry, which have long been less extensive than those enjoyed by its non-U.S. counterparts, have become considerably weaker in recent decades. It used to be that U.S. manufacturers and vendors supplied a significant part of the engineering information used to design and to build. They provided interface information for application and many actual details, such as equipment foundations or installation engineering. They also provided complete subsystems. The growing trend toward buying foreign, buying at lowest first cost, and buying components rather than systems, has weakened the linkages between construction companies and their material and equipment vendors. In addition, as components get larger and regulatory requirements become more complex, manufacturers are becoming increasingly unwilling or unable to accept the responsibilities they previously did, thus further weakening their ties to the construction firms. As more procurement goes overseas, and as fabrication and construction follow procurement, the associated engineering is likely to follow.

Contracting Practices

When housing and office buildings are excluded, public projects of governmental entities have accounted for over 50 percent of the total construction expenditures in recent years. The sheer size of public sector demand has meant that government contracting guidelines often serve as models for some sections of the private market. Under most government construction contracts, allowable (or recoverable) costs include direct costs of labor and associated fringe benefits, fixed or negotiated fees, and audited indirect costs such as rent, utilities, and equipment rental. Such contracts do not permit recovery of research and development costs and therefore discourage (if not actually prevent) investments in the industry's technological future. Since lowest estimated first cost is often the criteria for awarding contracts, even the use of the most experienced and most competent people is discouraged.

The Regulatory and Business Environment

Unlike their counterparts in many other countries, U.S. construction companies have neither protected markets nor public subsidies in the form of below market interest rates (patient money) or direct development support. When these handicaps are combined with a number of residual restrictive labor practices, more stringent antitrust regulations, and an aggressive liability legal system, technological change occurs very slowly and capital investment to accelerate such change is not readily available. All three factors (lack of vertical integration, contracting practice, unfavorable regulatory environment) discourage investment in long-term research and development for new technology acquisition in the United States and shorten the time horizons of U.S. construction firms.

National differences in management styles and managerial time horizons affect not only investment in R&D and new technologies but also the nature of technology flow between countries. For instance, U.S. engineering construction companies seem to be generally much more willing to transfer technology and know-how to foreign companies to seek short-term benefits of technology licensing or sales than vice versa. Moreover, even when foreign technology is made available to U.S. firms (even if the adoption of such technology would require zero capital investment by the recipient U.S. firm), they are slow in assimilating and using it. In short, while most U.S. construction firms currently discount the commercial relevance or importance of foreign construction technologies, they do so at their own peril. Over the next decade, the importance of foreign technical contributions to the technological vitality and competitiveness of the U.S. construction industry should grow dramatically. The combination of the development of management skills, the investment in technology development, and the higher level of inertia for change that exist in many U.S. companies means that their competitiveness will continue to decline unless the structural, regulatory, and attitudinal barriers to change in the U.S. market are dismantled.

One solution might be to develop a "product concept" for the engineering construction industry. If the item sold (and purchased) was a complete design, or a complete building or a bridge or a runway, the "cost" could then include the recovery of research and development costs as is done in most manufactured products, from automobiles to toasters to ballpoint pens. As long as the evaluation and bidding is based on manhours, this cannot be done. The product concept can encourage efficiency and provide an incentive to invest in technical advance and applications. How this product concept might be achieved and what might be the relative roles of government and industry is not obvious and warrants further discussion.

Overall, the construction engineering industry is similar to many other industries. The current emphasis in the United States on the short term, and the disincentives for U.S. firms to invest in the industry's technological future, will undermine their future competitiveness at home and abroad. Back on the farm, this practice is called eating the seed corn.

VII

Electrical Equipment and Power Systems Industry

WILLIS S. WHITE, JR.

This profile of the electrical equipment and power systems industry divides the subject into four pertinent categories: electrical, mechanical, nuclear, and innovative clean coal technology (see Table A-3). The profile was developed by listing the basic component parts of each category and identifying ongoing and emerging technologies and competencies.

An analysis of the profile clearly shows that U.S. electrical equipment manufacturing is in decline. This is also true of mechanical and nuclear system components sectors that make up the power system. The recent refusal of the U.S. government to allow the proposed turbine generator joint venture between Westinghouse and ASEA Brown Boveri highlights the problem. Electronic controls and monitoring along with developing clean coal technology are a few of the areas in which U.S. companies are holding their own. However, this apparent advantage in the area of clean coal technology may vanish if the proposed acid rain legislation passes in its present form.

The areas of decline in electric equipment and power systems industry consist of large and heavy equipment manufacturing such as transformers and circuit breakers, which are sometimes perceived as mature technologies. From the manufacturer's viewpoint, such equipment involves large capital commitments, high risks for successful completion, long lead times to manufacture, and relatively modest returns. Most of this equipment is tailor-made for the individual purchaser, and enough differences have been specified to make standardization difficult. Currently, research and development (R&D), as applied to this class of equipment, are used mainly for refining basic products.

For many years, corporate America worked hard at being the best it could be in its chosen field of endeavor. Its measure of success was domination of the market with its product reputation. This generally contributed to strong cash flows and modest steady returns. Of course, interest rates and inflation were also low. Plans were based on the long term, and even the individuals in those companies made long-term career commitments.

Today we hear discussions of "inner and outer circles." Companies exit businesses if they cannot be in first or second place in the market. Success is measured by a company's stock price. Corporate loyalty fades as individuals are shunted to and fro by restructuring, mergers, sell-offs, and dissolutions. Product R&D funding evaporates as these funds go toward improving the next quarter's profits. Management is geared for the short term.

This short-term syndrome of some U.S. electrical equipment manufacturers was typified in a *Wall Street Journal* article in which reporter Gregory Stricharchuk (1990) wrote, "But for all its successes, some skeptics wonder whether Westinghouse has come to epitomize the short-term mentality in corporate America. In its race to achieve quick returns, it may have dumped businesses that global competitors with more patience will ultimately profit from."

U.S. manufacturers have built and sold their equipment primarily for the American market, which was so large that there was little incentive to go after business overseas. The Europeans and the Japanese protected their home markets, and through organizations such as the International Electrical Association, coordinated their control of the world market. The U.S. government also did little to encourage overseas sales. Moreover, antitrust laws on the books from the late 1800s, to prevent American monopolies from dominating domestic business, effectively discouraged U.S. firms from collaborating on large international projects.

Foreign manufacturers have continued to grow and become stronger in the United States. They look to the American market for additional growth opportunities. Most foreign companies started doing business in the United States as small enterprises importing a few pieces of equipment here and there. As they became better acquainted with this market and their competition, they saw opportunities to buy small firms and in some cases to form joint ventures. For joint ventures, weaker companies that had a need of help were usually chosen. But it was not long before the foreign partner eventually took over the U.S. organization and created its own operating entity in the United States.

These foreign firms have in many cases maintained manufacturing in the United States. But, notable changes were emerging. Much of the heavy, low-tech work remains, while the high tech components are imported. Most of the engineering, particularly high-tech engineering, is completed overseas while the application engineering is completed locally.

A recent *Business Week* article stated that foreign companies have approximately a 25 percent share of the U.S. electrical equipment market.

If U.S. manufacturing loses its domestic market to foreign manufacturers, this automatically precludes its participation in the world market. Cohen and Zysman (1987) warn, "The U.S. and its companies must keep their mastery over manufacturing. You can't control what you can't produce."

Discussions with U.S. Department of Commerce officials have also shown that the government procurement code signed under the aegis of the General Agreement on Tariffs and Trade Treaty excludes electrical equipment from its jurisdiction. As a result, the U.S. market is open to all comers, while many foreign markets remain effectively closed to U.S. firms.

The responsibility for the reduction of U.S. manufacturing in the electrical equipment field cannot be placed solely on the manufacturer. Purchasers have some of the responsibility because they choose who will get the business. The basis of making an award generally comes down to the lowest price for which one can purchase the particular equipment. Intangibles such as reliability, service, and product quality are included in the evaluation. But items such as sources of future supply, the maintenance of technology in this country, the relationship of U.S. educational institutions to healthy manufacturers and their products, and the impact of the manufacturing jobs on the local economy are not often included in the evaluation.

In the case of nuclear power equipment manufacturing, the loss of U.S. manufacturing strength can also be tied to a lack of orders for new nuclear capacity. This is the consequence of uncertain regulation, visible political opposition, and a lack of resolve by elected government officials to maintain this energy option.

In summary, the reduction in U.S. electrical equipment manufacturing is the consequence of the present operating policies of three groups: manufacturers, users, and the U.S. government. The policies of these groups have been neither broad enough nor long-range enough to react to the emerging issues confronting the industry.

The problems faced by our country today are really nothing new in the fabric of time. In his 1926 book, *Today and Tomorrow*, Henry Ford discussed foreign competition. In Ford's view, overseas producers can undersell U.S. manufacturers only when domestic prices are "stupidly high." Then, he predicts, "competition would force the reorganization and replanning of these industries." As we enter into the 1990s, more than 60 years after Henry Ford's warning, much of American industry is being restructured.

REFERENCES

Cohen, Stephen S., and John Zysman. 1987. Manufacturing Matters: The Myth of the Post-Industrial Economy. New York: Basic Books.

Stricharchuk, Gregory. 1990. Westinghouse relies on ruthlessly rational pruning. Wall Street Journal. January 24.

TABLE A-3 Electrical Equipment and Power Systems Industry

Ongoing Technologies and Competencies	Emerging Technologies and Competencies
ELECTRICAL	
System Protection and Control	
Relays - U.S. has a dominant role but foreign supply and manufacturing ownership increasing.	Expertise moving offshore, increasing use of programmable logic controllers, computer and microprocessor-based relaying.
Line Trap and Coupling Capacitor Voltage Transformers (CCVTs) - Dominated by foreign manufacturers. Asea- Brown-Boveri's (ABB) purchase of Westinghouse's transmission and distribution business eliminates sole U.S. ownership.	Optical Potential Transformers and Current Transformers - limited U.S. participation.
Communication and Monitoring Equipment - U.S. dominated, many U.S. manufacturers.	Fiber optics - good U.S. base. Microprocessor-controlled monitoring equipment. Use of satellites to provide accurate timing for power disturbance monitoring.
Transformers	
Large (larger than 100 MVA) - One U.S.-owned supplier since ABB purchased Westinghouse R&D business.	Improved insulation systems -Nomex, Silicone - Both U.S. and foreign development.
Medium (40 to 100 MVA) - Rapidly decreasing number of U.S. suppliers. Increasing presence of foreign suppliers.	
Small (2.5 to 40 MVA) - Large number of U.S. suppliers and a healthy market.	
Circuit Breakers	
Extra High Voltage (above 240kV) - All foreign supplied.	Development of self-blast technology - Japanese lead, also European development.
High Voltage - One U.S. supplier.	Development of higher interrupting capability - Japanese lead, also European development.

Ongoing Technologies and Competencies	Emerging Technologies and Competencies
Medium and Low Voltage - Decreasing U.S. supply - Increasing foreign ownership.	Circuit breaker monitoring - Japan leads - some U.S. developments.
Surge Arresters	
Increasing foreign ownership.	Development of arrester materials with lower discharge voltage, better lifetime stability and higher energy capability - U.S. leads.
Motors	
Large (above 2500 HP) and Medium (to 2500 HP) - Increasing foreign manufacture.	Superconductivity - U.S., Europe, and Japan.
Small (1 to 200 HP) - Adequate U.S. suppliers with increasing U.S. manufactured foreign-owned motors.	Variable-speed motors - U.S., Europe, Japan.
Switches	
U.S. dominates manufacture.	
Gas Insulated Substations	
All foreign supplied and manufactured capabilities, Europe, Japan.	Development of manufacturing techniques and materials to reduce equipment size with increasing capabilities, Europe, Japan.
High Voltage Direct Current (HVDC)	
No U.S. maufacturers. Decreasing number of foreign manufacturers, increasing business.	HVDC circuit breaker development. Higher rated equipment - Europe, Japan. Thyrister technology - U.S., Europe, and Japan.
Wire and Cable	
High-voltage cable.	Foil barriers for waterproof cables - foreign owned.
115kV and above - extruded wire cable- one U.S. supplier, foreign manufacturers lead.	Fiber optics - U.S. patents, Japanese technology.
115kV pipe type - several U.S. manufacturers, foreign manufacturers lead.	
69kV - Several U.S. manufacturers.	
Medium Voltage (5 to 35kV) and Low Voltage (less than 5kV) - Primarily U.S. owned and manufactured.	

Table A-3 continues

Table A-3 continues

Ongoing Technologies and Competencies	Emerging Technologies and Competencies
Nuclear grade cable - significant decrease in number of vendors.	
Extrusion equipment - Majority is foreign produced.	
Transmission Towers Decreasing U.S. manufacturers, increasing foreign competition, with decreasing growth.	
MECHANICAL	
Major Pipe Supports and Hangers - Two domestic suppliers, no significant foreign ownership.	
Thermal Insulation - U.S. sources - small foreign presence.	R&D product development driven by environmental and health issues.
Valves - U.S. leads, foreign presence increasing, U.S. lags Japanese and European in casting technology.	Specialized control valve designs to improve operating life.
Heavy Wall Pipe and Pipe Fabricators - Very limited domestic production capability - Japan, Korea, and W. Germany increasing presence.	
Turbo Generators -	
Steam Turbines - Limited domestic suppliers, rapidly advancing foreign suppliers. U.S. losing technological advantage.	Superconducting generators - U.S., Japan, Europe.
Gas Turbines - Multiple domestic and foreign sources; Domestic manufacturing through GE and Westinghouse; Strong competitive market; Active R&D efforts by all manufacturers; Technology advancements held by all major manufacturers. Major R&D efforts in NO_X control and high efficiency combined cycles.	Ceramics, high-temperature blade coatings, high- strength, single crystal blade technology - U.S., European, Japanese all have a strong presence in this research.

Ongoing Technologies and Competencies	Emerging Technologies and Competencies
Steam Generators and Coal Pulverizer Equipment - Strong U.S. market share. Forming joint technology ventures with Japanese and European suppliers. U.S. lead but losing edge.	
Large Centrifugal and Axial Fans - Consolidation of U.S. suppliers. Foreign entry through domestic company purchase. Matured technology.	
Pumps - Major reduction in U.S. suppliers. European companies increasing their presence.	High-speed (15,000-20,000 rpm) pumps. Only one U.S. firm in R&D.
Feedwater Heaters - Adequate U.S. sources.	
Cooling Towers - U.S. supply adequate.	
Condensers - Through reduction in suppliers, adequate U.S. presence. Material-product-no current R&D effort.	
Precipitators - Increasing foreign presence.	Higher operating voltages (80-100kV) European technology.
Flue Gas Desulfurization System - Weak market, reduced U.S. suppliers. Increasing foreign supply capability.	Advanced chemistry and material applications - U.S. leads.
Instrumentation and Control - Strong U.S. presence.	Artifical intelligence - U.S. leads.
Caustic/Chlorine - U.S. dominates.	
NUCLEAR	
Products Used in Nuclear Plants but also Found in Fossil Plants - Nuclear qualification requirements becoming more expensive to obtain. See listing individual items above under Electrical and Mechanical.	

Table A-3 continues

Table A-3 continues

Ongoing Technologies and Competencies	Emerging Technologies and Competencies
Nuclear Fuel Assemblies and Related Components - U.S. sources dominant, but some foreign ownership.	U.S. showing strong leadership in light water reactor (LWR) fuel innovation.
	Ceramic-coated Particle Fuel Design for Gas Cooled Reactors - Lead shared by U.S. and West Germany.
	Graphite Fabrication - U.S. development equal to competition in U.K. and West Germany.
Reactor Pressure Vessels and Reactor Internals - No U.S. production - Current production in France, Japan, and U.K.	Prestressed Concrete Reactor Vessel. Leadership shared between Sweden, West Germany, and U.S.
	Liquid Metal Technology for Fast Breeder Reactors - France leads. Japan making a committed effort. Some U.S. activity.
Steam Generator Fabrication - Small U.S. replacement market controlled by Westinghouse. New plants in France, Japan, and U.K.	New material development (I690 Steam Generator Tubes) - high U.S. involvement.
Containment Construction - No current U.S. activity. Some activity in France, Japan and U.K.	
Nuclear Fuel Handling and Storage Equipment - Large number of U.S. sources.	
Uranium Conversion - U.S. maintains capability with increased Canadian participation.	
Uranium Enrichment - U.S. DOE retains most domestic business but DOE facilities are in trouble.	Laser enrichment technology for Uranium Enrichment - U.S. maintains lead.
Reactor System Design - No new reactors being constructed in the U.S. U.S. DOE is funding GE and Westinghouse to develop LWR designs using the natural laws of physics to accomplish reactor safety functions.	Helium circulators - Most experience in West Germany - some in U.S.
	Thermal Barrier "Density Locks"- Sweden leads, some R&D in U.S.

Ongoing Technologies and Competencies	Emerging Technologies and Competencies
Reactor Fuel Reprocessing and Plutonium Recovery - No U.S. activity.	France and U.K. are world leaders. Japan has strong effort.
Spent Fuel Disposal - U.S. effort proceeding slowly.	
INNOVATIVE CLEAN COAL TECHNOLOGIES	
Precombustion Cleaning (Advanced Coal Cleaning)	If and when advanced enough, U.S. companies will have major share of the market.
Will be dominated by U.S.-owned companies in the foreseeable future.	
During-Combustion Cleaning	
Fluidized-Bed Combustion	
— Atmospheric Bed	
When commercially available, over 80% will be dominated by U.S. suppliers/manufacturers.	U.S. will have competitive edge in developing more sophisticated I&C systems. Strong emerging European technology in circulating fluid beds.
— Pressurized Bed (PFBC)	
Combustor Assemblies: Presently envisioned that U.S. companies will serve domestic market.	European and Japanese companies are expected to provide significant competition in this area.
Boiler tubes: U.S. manufacturing capability declining. It is expected that the majority of tubing will come from foreign sources.	U.S. has taken the lead in focusing on in-bed tube wastage.
Cyclone/Hot Gas Cleanup: Both U.S. and foreign suppliers are expected to share the market.	Germany and Japan are spending considerable funds to develop an advanced hot gas cleanup system. As the development moves to high-tech, Westinghouse, Accurex Corporation, and other U.S. companies could influence the market, especially in the area of ceramic candle, ceramic cross-flow filters, etc.

Table A-3 continues

Table A-3 continues

Ongoing Technologies and Competencies	Emerging Technologies and Competencies
Coal preparation and injection system: Both U.S. and foreign suppliers and manufacturers share the market.	Improved and advanced systems may be dominated by foreign manufacturers.
Sorbent Feed System: Currently, both U.S. and foreign suppliers and manufacturers.	No emerging technologies are expected in this area.
Economizer: All boiler manufacturers in U.S. have capabilities of supplying this equipment.	With advancement of manufacturing technology, U.S. manufacturers would be more competitive. Foreign manufacturers (and especially Japan) may become more competitive with U.S.
Instrumentation and Control: Software-foreign suppliers. Hardware-Both U.S. and foreign suppliers and manufacturers.	This area will probably be dominated by U.S. suppliers after the maturity of the technology.
Valves and Piping: Mostly U.S. suppliers.	Expected to be dominated by U.S. suppliers.
Bed and Cyclone Ash Removal System: Current technology developed by foreign developers and manufacturers. U.S. has capability to enter this market when this technology matures.	Development of new technologies would put U.S. in a competitive market.
Gas Turbine: Currently only single foreign manufacturer. Market yet to be developed.	More U.S. manufacturers are expected to enter this market after the maturity of technology. However, U.S. manufacturers may not be able to compete in this area.
Slagging Combustors None	The technology has been developed in the U.S. as an after-growth of magnetohydrodynamic combustor development. All major suppliers are U.S.-owned. The market is expected to be dominated by U.S. companies, such as TRW, Rockwell, AVCO, and other conventional power plant equipment manufacturers.

Ongoing Technologies and Competencies	Emerging Technologies and Competencies
Post-Combustion Cleaning	
Induct Scrubbing: Presently all U.S. manufacturers.	U.S. could dominate this market.
Advanced Flue Gas Desulfurization System	
Most developers/suppliers are foreign-owned. However, some U.S. manufacturers under foreign licenses are willing to enter the market if the technologies could be applied with high- sulfur U.S. coal.	Development of high-tech manufacturing processes is not expected to change the market domination by foreign suppliers and manufacturers.
Coal Gasification Combined Cycle	
Over 85% U.S. suppliers and manufacturers, such as Texaco, Dow, Shell, Westinghouse, General Electric, M.W. Kellogg, etc. Only 15% of total will be supplied by West German, Swiss, and British suppliers. No real market has developed yet.	Development of high-tech could put U.S. in excellent shape to dominate the market. When the world market develops, the greatest proportion of that market is expected to be in the United States.
General Materials R&D	
(Basic Materials Research for all innovative clean coal technologies).	
Over 50% is dominated by Japan, Sweden, Switzerland, and West Germany.	Japan is expanding in this area to overtake the lead from U.S. and West Germany.

VIII

Semiconductor Industry

WILLIAM G. HOWARD, JR.

The semiconductor industry typifies many of the processes now driving the internationalization of engineering in many fields. The semiconductor business was recently dominated by U.S. technical efforts, but other countries are beginning to achieve technological parity (see TableA-4). Five major factors at work in this industry are as follows:

1. The semiconductor industry is seen to be one of the critical foundations for a national electronics industry, which in turn has been identified as a central focus by many countries seeking to develop industrial strength for the future. Semiconductors form the critical base for efforts in consumer electronics, computers, and communications hardware capability and support other related industrial efforts such as automobiles, aircraft, and robotics. Semiconductor competence also underlies much modern military hardware functionality for communications, avionics, guidance, radar, and electronic warfare weapons systems.

As such, virtually all industrially emerging nations have semiconductor industry development strategies. Those of Japan, Korea, Singapore, Hong Kong, Taiwan, and the People's Republic of China are noteworthy. Major efforts in the European Economic Community to strengthen semiconductor technology competence have also been mounted under the ESPRIT, RACE, Alvey, and Eureka programs.

The most aggressive strategies target not only the semiconductor device business, but the manufacturing and materials industries as well. These semiconductor strategies are designed to be stepping-stones to establishing more lucrative electronics hardware and systems businesses.

2. The U.S. semiconductor industry, despite its commanding global lead during the 1960s and 1970s, is vulnerable to international competition. Unlike its counterpart in several European and Asian Pacific countries, the U.S. industry has little vertical component. Each tier of the U.S. industry, from manufacturing equipment suppliers and materials vendors, to semiconductor device makers, to the primary semiconductor product users is made up of separate corporate entities, each dependent upon realizing a return on investment at their own point in the supply chain. With the exception of two or three captive lines, there is no mechanism whereby benefits realized at the system level are translated to priorities at the device, materials, or equipment levels.

The retarded development of the gallium arsenide device business in the United States as compared with the leadership achieved in Japan, particularly by Fujitsu and NEC, is a reflection of capability in the two countries to translate system-level needs into component business priorities. Furthermore, close working relationships between materials suppliers and device makers within Japanese company groups has significantly helped develop materials suppliers' technology.

U.S. companies, particularly in the manufacturing equipment area, tend to be small firms with little staying power when it comes to battling in the global marketplace against major, diversified company groups. This has led to serious loss of U.S. manufacturing and technology leadership, especially in parts of the industry concerned with fabrication materials, manufacturing equipment, dynamic memory, and consumer electronics components.

3. During the 1960s, the U.S. industry moved much of its labor-intensive manufacturing offshore to take advantage of lower labor costs and to gain access to foreign markets. Other international semiconductor manufacturers, particularly the Japanese, did the same but had strong incentives to find economic ways to repatriate manufacturing back into the home country in the 1970s. As a result, the Japanese tackled the problem of low-cost, automated manufacturing in an environment of high labor costs, while U.S. merchant manufacturers continued to move more activities to lower cost areas abroad. Virtually all volume assembly of semiconductors is now performed outside the United States, and technical control of those activities has followed.

As offshore manufacturing activities increased, critical engineering and technical support activities followed in order to remain in close proximity to factories and foreign customers. Engineering activities were staffed with foreign nationals, who now represent the core competence in a number of critical technical areas in some major U.S. firms.

4. As the semiconductor industry has matured, the technology has spread across the globe. The process started with U.S. multinational corporation-

trained foreign engineers, spread to U.S. university-educated scientists and engineers returning to their home countries to work in local firms or as semiconductor users, and has achieved critical mass with the establishment of competent semiconductor and solid-state physics engineering programs in universities worldwide. Possession of the technology is no longer unique, and the open, international technical publication and conference system helps sustain the universal understanding of many of the latest developments. In the semiconductor industry, the genie is out of the bottle but, realistically, could never have been confined in the long term.

The recent success of major Korean companies at purchasing and adapting the technical know-how with which to start up several semiconductor producers demonstrates how freely the technology, materials, and manufacturing equipment flow worldwide.

5. The semiconductor technology continues to evolve rapidly. With each major change, the established patterns of competition in the industry are subject to upset. This vulnerability has been evident at major turning points in semiconductor technology:

- Vacuum tubes to discrete transistors
- Discrete transistors to integrated circuits
- Small Scale Integrated (SSI) circuits and Medium Scale Integrated (MSI) circuits to microprocessors and memories
- Standard, high-volume commodity products to application-specific products

At each of these technologically driven transitions, new entrants have displaced older, less adaptive companies in the fastest growth segments of the business. Similar dynamic processes have been at work in the materials and manufacturing equipment portions of the semiconductor industry. Technological changes have provided opportunities for new entrants at every level of the business to compete on an equal footing with more established current leaders.

Each of these five forces can be seen at work in other industries as well. However, the rapidity with which they have made major shifts in the international engineering balance is striking in the semiconductor case.

TABLE A-4 Semiconductor Industry Technology Profile

Ongoing Technologies and Competencies	Emerging Technologies and Competencies
1. *Lithography/optics* Foreign leadership, U.S. sources flagging, foreign control of lens supply.	1. *Galium Arsenide* Japanese lead, U.S. users turn to Japanese suppliers.
2. *Fabrication equipment* Japanese control, U.S. lags with some exceptions.	2. *Molecular beam epitaxy (MBE)/Metallo-organic oxidative chemical vapor deposition (MOCVD)* U.S. lead MBE, Japan lead MOCVD.
3. *Design* U.S. lead.	3. *X-ray lithography* Japanese lead.
4. *Computer-aided design/ Computer-aided manufacturing* U.S. lead. U.S. suppliers sell to all comers.	4. *Engineered materials* U.S. lead.
5. *General materials/ceramics* Crystal silicon: 2 German, 4 Japanese firms dominate, most U.S. sourcing offshore, U.S. has lost this capability.	5. *Electron beam lithography* JEOL/Cambridge (Japan/UK) lead.
6. Manufacturing skills Automated equipment, materials. U.S. lag.	
7. Diffusion implant U.S. lead, but sell to all comers.	

APPENDIX

B

Contributors

The committee acknowledges the valuable insight and contributions made by the following people and organizations during meetings or through correspondence.

PIERRE R. AIGRAIN (France), Scientific Advisor to the President, Thomson Group

HORST ALBACH (Federal Republic of Germany), President, Berlin Academy of Sciences and Technology

PHILIP G. ALTBACH (U.S.A.), Professor, State University of New York at Buffalo

JOHN A. ARMSTRONG (U.S.A.), Vice President for Science and Technology, IBM Corporation

PETER-MICHAEL ASAM (Federal Republic of Germany), Corporate Research and Technology, Siemens AG

GEOFFREY ATKINSON (United Kingdom), Executive Secretary, The Fellowship of Engineering

JESSE H. AUSUBEL (U.S.A.), Fellow in Science and Public Policy, The Rockefeller University; Director of Studies, Carnegie Commission on Science, Technology, and Government

LIONEL O. BARTHOLD (U.S.A.), Chairman and Principal Consultant, Power Technologies, Inc.

ALDEN BEAN (U.S.A.), Professor and Director, Center for Innovative Management Studies, Lehigh University

STEPHEN D. BECHTEL, JR. (U.S.A.), Chairman, Bechtel Group

J. C. BELOTE (U.S.A.), Vice President, Bechtel Group

JENNIFER SUE BOND (U.S.A.), Study Director, International Studies Group, Division of Science Resources Studies, National Science Foundation

MICHAEL BORRUS (U.S.A.), BRIE, University of California, Berkeley

ROBERT BRAINARD (France), Principal Administrator, Directorate for Science, Technology and Industry, Organization for Economic Cooperation and Development

HARVEY BROOKS (U.S.A.), Professor, John F. Kennedy School of Government, Harvard University

GEORGE BUGLIARELLO (U.S.A.), President, Polytechnic University

K. BULTHUIS (The Netherlands), Senior Managing Director, Philips International B.V.

JOHN P. CAMPBELL (U.S.A.), Senior Program Officer, Government-University-Industry Research Roundtable for the National Academy of Sciences, National Academy of Engineering, Institute of Medicine.

FRANK CARRUBBA (U.S.A.), Director, Hewlett-Packard Laboratories, Hewlett-Packard Company

ARTHUR Y. CHEN (Taiwan), President, Ret-Ser Engineering Agency

DONALD A. CHISOLM (Canada), Retired Executive Vice President, Technology and Innovation, Northern Telecom, Ltd.

KUN MO CHUNG (Korea), President, Korea Science and Engineering Foundation

DAVID E. COLE (U.S.A.), Director, Office for the Study of Automotive Transportation, Transportation Research Institute, The University of Michigan

UMBERTO COLUMBO (Italy), Chairman, ENEA, Italian National Agency for R&D Nuclear and Alternative Energy Resource

PHILIP M. CONDIT (U.S.A.), Executive Vice President, The Boeing Company

RHONDA J. CRANE (U.S.A.), Senior Advisor for Science and Technology, Office of the United States Trade Representative

MICHAEL CUSUMANO (U.S.A.), Professor, Sloan School of Management, Massachusetts Institute of Technology

CARL DAHLMAN (U.S.A.), Industry Development Division, World Bank

DONALD A. DAHLSTROM (U.S.A.), Research Professor, Chemical Engineering Department, University of Utah

H. G. DANIELMEYER (Federal Republic of Germany), Direktor, Corporate Research and Development, Siemens AG

RICHARD D. DELAUER (U.S.A.), Chairman, The Orion Group Ltd. — deceased

THOMAS F. DONOHUE (U.S.A.), General Manager of Advanced Technology Operations, GE Aircraft Engines

SIR DIARMUID DOWNS (United Kingdom), Consultant and Former Chairman, Ricardo Consulting Engineers PLC, Bridge Works

JOHN L. DOYLE (U.S.A.), Executive Vice President, Hewlett-Packard Co.

JOHN M. DUTY, JR. (U.S.A.), Vice President and Manager of Engineering, Bechtel Group

SIGVARD EKLUND (Austria), Director General Emeritus, International Atomic Energy Agency

RICHARD ELLIS (U.S.A.), Engineering Manpower Commission, American Association of Engineering Societies

WALTER ENGL (Federal Republic of Germany), Institute für Theoretischer Elektrotechnik, Rheinische-Westf., Technische Universität

LEO ESAKI (U.S.A.), Fellow, T. J. Watson Research Center, IBM

ALAN FECHTER (U.S.A.), Executive Director, Office of Scientific and Engineering Personnel, National Research Council

MICHAEL FINN (U.S.A.), Director, Studies and Surveys, Executive Director's Office, Office of Scientific and Engineering Personnel, National Research Council

JOHN W. FISHER (U.S.A.), Professor and Director, ATLSS Engineering Research Center, Lehigh University

HANS FORSBERG (Sweden), President, Royal Swedish Academy of Engineering Sciences

STUART M. FREY (U.S.A.), Vice President, Corporate Quality and Technical Affairs, Ford Motor Company

ALBERT J. GRIMARD (U.S.A.), Engineering Operations Section Head, Boston Edison

GEORGE H. HEILMEIER (U.S.A.), Senior Vice President and Chief Technical Officer, Texas Instruments

BACHARUDDIN J. HABIBIE (Indonesia), Minister of State for Research and Technology

RICHARD A. HACKBORN (U.S.A.), Vice President and General Manager, Peripherals Group, Hewlett-Packard

WOLF HAFELE (Federal Republic of Germany), Director General, Kernforschungsanlage Juelich GmbH

JOHN HAGEDOORN (The Netherlands), Senior Research Fellow, Maastricht Economic Research Institute of Innovation and Technology (MERIT)

YOSHINORI HARAGUCHI (Japan), Senior Researcher, R&D Planning Department, Tokyo Electric Power Company

MARTHA C. HARRIS (U.S.A.), Director, Office of Japan Affairs, National Research Council

JEFFREY A. HART (U.S.A.), Berkeley Roundtable on the International Economy (BRIE), University of California, Berkeley

ABRAR HASAN (France), Principal Administrator, Central Analysis Division, Organisation of Economic Cooperation and Development

YOSHIHIDE HASE (Japan), General Manager, Power Systems Division, Toshiba Corporation

HERMANN HAUSER (United Kingdom), Vice President, Research, Active Book Company

ROBERT HAYES (U.S.A.), Professor, Harvard Business School
KEN HEYDON (France), Principal Administrator, Trade Directorate, Organization for Economic Cooperation and Development
CHRISTOPHER HILL (U.S.A.), Executive Director, Manufacturing Forum, National Academy of Engineering and National Academy of Sciences
KEIICHI HORI (Japan), Deputy General Manager, Non-Ferrous New Materials and Semi-Products Department, Sumitomo Corporation
MELVIN HORWITCH (U.S.A.), Sloan School of Management, Massachusetts Institute of Technology
MICHEL HUG (France), C.T.C.
HIROSHI INOSE (Japan), Director General, National Center for Science Information System
YOKICHI ITOH (Japan), Managing Director, Corporate Research Laboratories, Fuji Xerox Company, Ltd.
ISAO IYODA (Japan), Power Systems Section, Energy and Industry Engineering Department, Mitsubishi Electric Corporation
THOMAS J. JUDGE (U.S.A.), President and Chief Executive Officer, The Austin Company
KATHRYN JACKSON (U.S.A.), NAE Fellow, Program Office, National Academy of Engineering
ANDREW L. JACOB (U.S.A.), Secretary, Association of Electric Illuminating Companies
GUNTER JAENSCH (Federal Republic of Germany), Corporate Research and Development, Siemens AG
ERNEST G. JAWORSKI (U.S.A.), Director of Biological Sciences, Monsanto Company
YONGWOOK JUN (Korea), Assistant Professor, College of Business Administration, Chung-Ang University
WILLIAM L. KATH (Japan), Senior Managing Director, Mazda Motor Corporation
MAKATO KIKUCHI (Japan), Managing Director, Sony Research Center, Sony Corporation
YUJI KIMURA (Japan), Nuclear and Electric Power Division, Toyo Engineering Corporation
PETER KIRSTEIN (United Kingdom), Chairman, Computer Science Department, University of London
HERWIG KOGELNIK (U.S.A.), Director, Photonics Research Laboratory, AT&T Bell Laboratories
SHIRO KURIHARA (Japan), Director for Development Program, Moonlight Project, AIST, Ministry of International Trade and Industry
KANEYUKI KUROKAWA (Japan), Managing Director, Fujitsu Laboratories, Ltd.
CHONG OUK LEE (Korea), Center for Science and Technology Policy, Korea Institute of Science and Technology

KANGHI CELESTE LEE (Korea), Executive Director, Korea Business World
KEI-HONG LEE (Korea), Publisher, Korea Business World
WILLIAM S. LEE (U.S.A.), Chairman and President, Duke Power Company
RICHARD K. LESTER (U.S.A.), Commission on Productivity, Massachusetts Institute of Technology
CHANG LIN-TIEN (U.S.A.), Chancellor, University of California, Berkeley
HANS LIST (Austria), Chairman, AVL Gmb.H
THOMAS C. MAHONEY (U.S.A.) Acting Executive Director, Manufacturing Studies Board, National Research Council
ROBERT MALPAS (United Kingdom), Managing Director, British Petroleum Company, PLC
CRAIG MARKS (U.S.A.), Vice President, Technology and Productivity Planning, Automotive Sector, Allied-Signal, Inc.
SHINJI MATSUMOTO (Japan), General Manager, Construction Engineering Research Department, Institute of Technology, Shimizu Corporation
E. R. McGRATH (U.S.A.), Chairman, Association of Electric Illuminating Companies
W. COURTNEY McGREGOR (U.S.A.), Vice President, Technical Development, Xoma Corporation
AKIO MITSUFUJI (Japan), Project Manager, Industrial Systems Engineering Division, Toyo Engineering Corporation
TSUNEO MITSUI (Japan), Managing Director, The Tokyo Electric Power Company, Inc.
HEE MOCK NOH (Korea), Director, Lucky-Goldstar Economic Research Institute
DAVID C. MOWERY (U.S.A.), Professor, University of California, Berkeley
PETER MULLER-STOY (Federal Republic of Germany), Corporate Research and Technology, Siemens AG
KANJI MURASHIMA (Japan), Associate Board Director, Vice President, Aero-Engine and Space Operations, Ishikawajima-Harima Heavy Industries, Co., Inc.
RYOICHI NAKAGAWA (Japan), Senior Technical Advisor, Central Engineering Labs, Nissan Motor Company, Ltd.
NEBOJSA NAKICENOVIC (Austria), Technology, Economy and Society Program, International Institute for Applied Systems Analysis
KEE DONG NAM (Korea), Vice Chairman, Tong Yang Cement Corporation
JURGEN NAUJOKS (Federal Republic of Germany), Corporate Research and Technology, Siemens AG
RICHARD NELSON (U.S.A.), Professor, School of International and Public Affairs, Columbia University
TOSHIO NODA (Japan), Corporate Mangement Advisor, Osaka Titanium Company, Ltd.
JUNJI NOGUCHI (Japan), Executive Director, Union of Japanese Scientists and Engineers

AKIO NUMAZAWA (Japan), Executive Vice President, Tokai Rida Company, Ltd.
BRIAN OAKLEY (United Kingdom), Chairman, Logica Cambridge Limited
MICHEL ODELGA (United Kingdom), Director, International Trade Relations, Rank Xerox Ltd.
AKIO OHJI (Japan), Senior Manager, Turbine Engineering, Turbine Plant, Planning Department, Toshiba Corporation
SHOZO OJIMI (Japan), Board Director of IHI, President of Operations for Aeroengine and Space Operations, Ishikawajima-Harima-Heavy Industries Company Ltd.
TAKASHI OKABE (Japan), Executive Director, Nippondenso
JAMES Y. OLDSHUE (U.S.A.), Vice President, Mixing Technology, Mixing Equipment Company, Inc. - Unit of General Signal
CARLOS S. OSPINA (Colombia), Senior Partner and Manager, INGETEC
WON HEE PARK (Korea), President, Korea Institute of Science and Technology
ARNO PENZIAS (U.S.A.), Vice President, Research, AT&T Bell Laboratories
LOIS PETERS (U.S.A.), School of Management, Lally Management Center, Rensselaer Polytechnic Institute
CORRADO PIRZIO-BIROLI (U.S.A.), Deputy Head of Delegation, European Communities Delegation
JOHN H. PROVANZANA (U.S.A.) Manager, Major Transmission Equipment, American Electric Power Company, Inc.
ROBERT PRY (Austria), Director, International Institute for Applied Systems Analysis
MIKE R. REECE (United Kingdom), GEC Hirst Research Centre
FRANK W. RIES (U.S.A.), Chairman, Stone & Webster Engineering Corporation
GUSTAVO RIVAS-MIJARES (Venezuela), Professor, Central University of Venezuela
SIR DENIS ROOKE (United Kingdom), Chairman, British Gas plc
DANIEL ROOS (U.S.A.), Professor, Technology and Public Policy Program, Massachusetts Institute of Technology
RICHARD ROSENBLOOM (U.S.A.), Professor, Harvard Business School, Harvard University
JACQUES ROSSIGNOL (France), Vice President of Engineering, SNECMA
SHIOCHI SABA (Japan), Executive Advisor, Toshiba Corporations
AKIO SAKURAI (Japan), Director, Planning Division, CRIEPI
MASAYUKI SAKURAI (Japan), Group Executive, Technology Information and Communication Group, Toshiba Corporation
HAROLD N. SCHERER, JR. (U.S.A), Senior Vice President, Electrical Engineering, American Electric Power Company, Inc.
HEINZ SCHWAERTZEL (Federal Republic of Germany), Corporate Research and Technology, Siemens AG

YASUJI SEKINE (Japan), Professor, Department of Electrical Engineering, University of Tokyo

PASCAL SENECHAL (France), Director of Technology, SNECMA

SATOSHI SHINOZAKI (Japan), Senior Manager, Advanced Memory Technology Department, Semiconductor Device Engineering Laboratory, Toshiba Corporation

DENIS FRED SIMON (U.S.A), Associate Professor of International Business Relations, The Fletcher School of Law and Diplomacy, Tufts University

CLAUDIA STAINDL (Austria), International Institute for Applied Systems Analysis

HELENA STALSON (U.S.A.), Research Consultant

BARRY STEVENS (France), Planning and Evaluation Unit, Office of the Secretary General, Organization for Economic Cooperation and Development

CANDICE STEVENS (France), Directorate for Science, Technology and Industry, Organization for Economic Cooperation and Development

NORIHASA SUZUKI (Japan) Director, IBM Tokyo Research Laboratories

SEISHI SUZUKI (Japan), General Manager, Engineering Department, Shimizu Group, Shimizu Construction Company

KOICHI TAKIGUCHI (Japan), Deputy Senior Staff Manager, Xerographic Technology Research Laboratory, Fuji Xerox Co., Ltd.

TAKAHITO TANABE (Japan), General Manager, Overseas Planning and Operations Department, Tokai Rika co., Ltd.

TARO TANAKA (Japan), President, Nippondenso Co., Ltd.

TAKESHI TASHIRO (Japan), Manager, Business Development, Aero-Engine and Space Operations, Ishikawajima-Harima Heavy Industries Co., Ltd.

JOHN TAYLOR (United Kingdom), Director, Bristol Research Center, Hewlett-Packard

MYRON B. TRENNE (U.S.A.), General Manager, Corporate Research and Development, Eaton Corporation

MASAMI UEDA (Japan), Engineering R&D Administration, Tokyo Electric Power Co., Inc.

MICHIYUKI UENOHARA (Japan), Executive Advisor, NEC Corporation

YOICHI UNNO (Japan), General Manager, Semiconductor Device Engineering Laboratory, Toshiba Corporation

GEORGES ANDRE CHARLES VENDRYES (France), Scientific Advisor to the Chairman, Atomic Energy Commission of France

RAYMOND VERNON (U.S.A.), Professor Emeritus, John F. Kennedy School of Government, Harvard University

SHELDON WEINIG (U.S.A.), Chairman and CEO, Materials Research Corporation

GUNNAR WESTHOLM (France), Administrator, Scientific, Technological and Industrial Indicators Division, Directorate for Science, Technology and Industry, Organization for Economic Cooperation and Development

STUART R. WETTERSCHNEIDER (U.S.A.), Vice President, Design & Engineering, The Austin Company

CLAUS WEYRICH (Federal Republic of Germany), Corporate Research and Development, Siemens AG

DAVID WHEAT (U.S.A.), Vice President, The Boston Capital Group

MAURICE V. WILKES (England), Olivetti Research Ltd.

F. KARL WILLENBROCK (U.S.A), Assistant Director, Science, Technology and International Affairs, National Science Foundation

THOMAS E. WILSON (U.S.A.), Manager GE90 Business Requirement, GE Aircraft Engines

PATRICK WINDHAM (U.S.A.), Professional staff member, Committee on Commerce, Science and Transportation, U.S. Senate

JAMES WOMACK (U.S.A.), Massachusetts Institute of Technology

GREGORY K. WURZBURG (France), Principal Administrator, Directorate for Social Affairs, Manpower and Education, Organization for Economic Cooperation and Development

TUNIHIKO YOKOYAMA (Japan), Manager, Secretary's Office, Toshiba Corporation

TADAO YOSHIDA (Japan), Director, General Manager—Nuclear and Electric Power Department, Toyo Engineering Corporation

SHIGEKAZU YOSHIJIMA (Japan), Group Executive, Technology Energy Systems Group, Toshiba Corporation

EIICHI ZAIMA (Japan), Senior Research Engineer, Engineering Research Center, Tokyo Electric Power Company

JOSEPH F. ZIOMEK (U.S.A.), TRW Vehicle Safety Systems Inc.

APPENDIX

C

Biographical Information on Committee Members

THOMAS D. BARROW, the former chairman of Kennecott Copper Company, was elected vice chairman of the Standard Oil Company Ohio, now B.P. America in 1981. He managed Sohio's oil and natural gas exploration and production activities plus the worldwide minerals business of Kennecott Corporation until his retirement in 1985. His career as a geologist began in 1951 at Humble Oil and Refining Company (Exxon) where he served in various capacities, including president. He later became senior vice president of Exxon Corporation and member of the board (1972–1978). His responsibilities covered worldwide exploration and production activities, mining and synthetic fuels, science and technology, and corporate planning. Dr. Barrow is a member of the National Academy of Engineering, a trustee of Stanford University and Baylor College of Medicine, and a former trustee of Woods Hole Oceanographic Institution and the American Museum of Natural History. He received his Ph.D. in geology from Stanford University.

W. DALE COMPTON is Lillian M. Gilbreth Distinguished Professor of Industrial Engineering at Purdue University. Prior to his appointment at Purdue in 1988, he served as senior fellow at the National Academy of Engineering (1987–1988). In 1973, after three years service as director of the Ford Motor Company's chemical and physical science division, Dr. Compton was appointed Ford's vice president for research, a position he held until 1987. Before joining Ford, he was professor of physics and director of the Coordinated Science Lab at the University of Illinois. Dr. Compton has worked as a consultant for both private and federal research organizations. He is a fellow of the American Physical Society and the American Association for the Advancement of Science, and a member of the National Academy of Engineering. Dr. Compton received his Ph.D. in physics from the University of Illinois.

ELMER L. GADEN, JR. has served as Wills Johnson Professor of Chemical Engineering at the University of Virginia since 1979. Before joining the faculty at the University of Virginia, Dr. Gaden was dean of the College of Engineering, Mathematics and Business Administration at the University of Vermont (1975–1979), and a faculty member at Columbia University (1949–1974), where he taught chemical engineering, bioengineering, and history. Dr. Gaden was the founding editor of the international research journal *Biotechnology and Bioengineering* and served as editor for 25 years (1959–1983). He is a member of the National Academy of Engineering. Dr. Gaden received his Ph.D. in chemical engineering from Columbia University.

DONALD L. HAMMOND recently retired as director of Hewlett-Packard Laboratories after 29 years at the company. From 1983 to 1986 he started the HP European Research Center in Bristol, England, focused on computer science and data communication. As one of the founders of HP Laboratories in 1966, he directed the Physical Electronics Laboratory and the Physics Research Center. He managed production and development of quartz crystal devices at Hewlett-Packard and scientific electronic products. He is a member of the National Academy of Engineering and a fellow of the Institute of Electrical and Electronics Engineers. His background is in physics with B.S. and M.S. degrees from Colorado State University. He has received honorary doctorates from the University of Bristol and Colorado State University.

WILLIAM G. HOWARD, JR. is a senior fellow at the National Academy of Engineering, currently on leave from Motorola, Inc. where he served most recently as senior vice president and director of research and development. His focus at the Academy is in the area of technology commercialization in private industry. He has served on numerous government and private advisory panels and has served as chairman of the U.S. Department of Commerce's Semiconductor Technology Advisory Committee and currently chairs a working group of the Department of Defense's advisory group on electron devices. Before joining Motorola in 1969, Dr. Howard was an assistant professor of electrical engineering and computer sciences at the University of California, Berkeley, where he earned his doctorate. He is a member of the National Academy of Engineering and has held a variety of positions in the Institute of Electrical and Electronics Engineers.

TREVOR O. JONES is chairman of the board of Libbey-Owens-Ford Company and also president of the International Development Corporation (IDC) of Cleveland, Ohio. A native of Maidstone, England, Mr. Jones started his U.S. engineering career with General Motors in 1959, where he spent 19 years working in aerospace activities and in 1970 was charged with bringing aerospace technology to automotive safety and electronic systems. He became director of GM's newly organized Automotive Electronic Control Systems group in 1970, was appointed director of Advance Product Engineering in 1972, and became director of GM's Proving Grounds in 1974. Mr. Jones was employed by TRW in a number of executive positions, including vice president of engineering TRW Automotive Worldwide, group vice president and general manager of TRW's Transportation Electrical and Electronics Group, and group vice president, Stategic Planning, Business Development, and Marketing for the Automotive Sector.

He is a fellow of the British Institute of Electrical Engineers, the American Institute of Electrical and Electronics Engineers, the Society of Automotive Engineers, Inc., and a member of the National Academy of Engineering. He has received many awards for his work in automobile electronics and safety and has been cited many times for his leadership in the application of electronics to the automobile. Mr. Jones completed his formal engineering education in the United Kingdom at Aston Technical College and Liverpool Technical College.

THOMAS H. LEE is professor emeritus of electrical engineering at the Massachusetts Institute of Technology and president of the Center for Quality Management. In 1948 he began work with General Electric where, over the course of 32 years, he held numerous posts from senior research engineer (1955–1959) to staff executive and chief technologist (1978–1980). In 1980 he left General Electric to become director of the Electric Power Systems Engineering Laboratory and Philip Sporn Professor of Energy Processing at the Massachusetts Institute of Technology. In 1984 he became director of the International Institute for Applied Systems Analysis in Laxenburg, Austria, for a three-year term. He is a member of the National Academy of Engineering. Dr. Lee received his doctorate in electrical engineering from Rensselaer Polytechnic Institute.

MILTON LEVENSON, recently retired, was executive engineer and special assistant to the president at Bechtel Power Corporation since 1981. In 1943 he began work as junior engineer and has since worked at Oak Ridge National Laboratory, the U.S. Army Manhattan Engineering District, and held progressively advanced positions at Argonne National Laboratory, ending as associate laboratory director of energy and environment. From 1973 to 1981, he was the director of the nuclear division at the Electric Power

Research Institute. He is past president of the American Nuclear Society, a member of the National Academy of Engineering, and has served on U.S. technical delegations to four Geneva conferences on peaceful use of atomic energy. Mr. Levenson received his bachelor's degree in chemical engineering from the University of Minneapolis and an MBA from the University of Chicago.

PETER W. LIKINS is president of Lehigh University. Dr. Likins received his Ph.D. in engineering mechanics and his bachelor's degree in civil engineering from Stanford University, with an intervening master's degree from the Massachusetts Institute of Technology and experience as a development engineer at the Jet Propulsion Laboratory of the California Institute of Technology. He served on the engineering faculty at the University of California, Los Angeles, from 1964 to 1976, when he became professor and dean at Columbia University. From 1980 to 1982 he was provost at Columbia; then he moved to his present position. He holds honorary degrees from Lafayette and Moravian Colleges and the Medical College of Pennsylvania. He is a fellow of the American Institute of Aeronautics and Astronautics, a member of the National Academy of Engineering, and a member of the U.S. President's Council of Advisors on Science and Technology.

EDWARD A. MASON, an independent consultant, was vice president of research at Amoco Corporation from 1977 until his recent retirement in 1989. He started his engineering career as assistant professor of chemical engineering at the Massachusetts Institute of Technology, became professor of nuclear engineering and head of the department of nuclear engineering in 1971. From 1953 to 1957 he was director of research at Ionics, Incorporated. During subsequent periodic leaves of absence from MIT, he worked at the Oak Ridge National Laboratory as a senior design engineer, the National Science Foundation's Euratom Research Center, and the U.S. Nuclear Regulatory Commission as commissioner from 1975 to 1977. He is a member of the National Academy of Engineering and of numerous professional societies, including the American Academy of Arts and Sciences, the American Association for the Advancement of Science, the American Chemical Society, the American Insitute of Chemical Engineers, and the Industrial Research Institute. He received his Sc.D. from the Massachusetts Institute of Technology.

BRIAN H. ROWE has served as senior vice president, GE Aircraft Engines, with General Electric Company since 1979. He started his career with GE in 1957, after having worked at the deHavilland Engine Company in England. He has worked in design engineering, marketing, engineering production, and has held senior management positions in commercial airline and aircraft engine engineering divisions. He is a member of the National Academy of Engineering, the American Institute of Aeronautics and Astronautics, and a fellow of the Royal Aeronautical Society of England. He holds seven patents. Mr. Rowe received his degree in mechanical engineering from Kings College, Durham University in England.

WILLIAM J. SPENCER became president and chief executive officer of Sematech in November 1990, having served as group vice president for corporate research at Xerox Corporation since 1986. Before he joined Xerox in 1981, Dr. Spencer held senior management positions in R&D at AT&T. His interests include the management of technology, innovation, global industries, and engineering education. A member of the National Academy of Engineering, he has served on engineering advisory panels at Columbia University, the University of California at Berkeley, Stanford University, the University of Illinois, and Princeton University. Dr. Spencer received his Ph.D. in physics from Kansas State University.

WILLIS S. WHITE, JR. has been with the American Electric Power System since graduation from college in 1948. In 1976 he became chairman of the board of American Electric Power Company and its chief executive officer. He is also chairman and chief executive officer of each of AEP's operating companies and subsidiaries, and is president of Ohio Valley Electric Corporation, serving the U.S. Department of Energy. Mr. White is chairman of the Ohio Center's board of trustees, trustee at Battelle Memorial Institute, and director for the Bank of New York. He is a member of the National Academy of Engineering. Mr. White is an electrical engineering graduate of Virginia Polytechnic Institute and State University and holder of a master's degree in industrial management from the Massachusetts Institute of Technology.

Index

C

D

E

I

J

K

L

M

N

O

P